PRAISE FOR

Rivals

"In her graceful sweep through four centuries of scientific collaboration, Lorraine Daston recounts how groups of scientists have gotten together to get things done."

DAVA SOBEL
Author of *Longitude, Galileo's Daughter,* and *The Glass Universe*

"No one can disentangle the genealogy of scientific values like Lorraine Daston. In this elegant book, she trains this admirable skill on the very idea of the international scientific community. Scientists and laypeople alike will find in *Rivals* a lively new way of thinking about how the cosmopolitan astronomy of the Enlightenment has given way to the global climate science of today."

KEN ALDER
Author of *The Measure of All Things*

"How, when, and why did the notion of science as a shared and ultimately a global endeavor emerge—and how is it faring in the ultracompetitive digital age? By exploring these questions with wit, verve, concision, perspicacity, and deep learning, Lorraine Daston has produced an essential resource for anyone interested in how science works and how it came to work that way."

PHILIP BALL
Author of *Curiosity: How Science Became Interested in Everything*

"Lorraine Daston is as familiar with the perennial tensions that make scientists behave like rivals, not partners, as she is knowledgeable about collaborations in science, even those buried deep in history. Her graceful, short, and compelling account of how and why large-scale scientific collaborations have occurred in the past, how they have thrived (especially without intrusions by governments), and why we need to promote them today is a story that should be read by anyone who does scientific work or benefits from it."

HAROLD VARMUS
Lewis Thomas University Professor,
Weill Cornell Medicine and Nobel Laureate
in Physiology or Medicine

Rivals
How Scientists Learned to Cooperate

COLUMBIA GLOBAL REPORTS
NEW YORK

Rivals

How Scientists Learned to Cooperate

Lorraine Daston

PARTICIPATING OBSERVATORIES
IN THE CARTE DU CIEL
Original locations
Later locations/replacements
Mexico
Tacubaya
Brazil
Rio de Janie
Santiago
Chile
La Plata
Argentina

Finland
Helsinki
United Kingdom
Belgium
Oxford
Uccle
Potsdam
Greenwich
Germany
Paris
France
Italy
Bordeaux
Toulouse
Vatican
Cordoba
Spain
Algiers
Catania
San Fernando
Algeria
India
Hyderabad
Australia
Perth
Sydney
Melbourne
Cape of Good Hope
South Africa
© 2023 Jeffrey L. Ward

For Gerd

Published with support from the Andrew W. Mellon Foundation

Based on the Menahem Stern Lectures, 2022

Published by Columbia Global Reports
91 Claremont Avenue, Suite 515
New York, NY 10027
globalreports.columbia.edu
facebook.com/columbiaglobalreports
@columbiaGR

Library of Congress Cataloging-in-Publication Data Available Upon Request

ISBN 979-8-9870535-6-0 (TP)
ISBN 979-8-9870535-7-7 (eBook)

Book design by Strick & Williams
Map design by Jeffrey L. Ward
Author photograph by SCAS (Swedish Collegium for Advanced Study)
(Front) Band of visibility of 1761 transit of Venus. Bibliothèque de l'Observatoire de Paris.
(Back) *Mammato-cumulus*, H. Hildebrandsson et al. (eds.), *Atlas international des nuages* (1896). Table XIII, Fig. 26.

Printed in the United States of America

CONTENTS

LIST OF FIGURES

The Uncommunal Community

The scientific community is by any measure a very strange kind of community. For starters, no one knows who exactly belongs to it, much less who speaks for it. Its members are a miscellany of individuals but also of disparate institutions: universities, research institutes, government agencies, international organizations, learned societies and journals, and now preprint servers and online data archives. Nor does it have a fixed location. Despite the cozy, *gemeinschaftlich* associations of the word "community," the village conjured up by the term "scientific community" is scattered all over the globe and its inhabitants meet only occasionally, if at all. Far from living in neighborly harmony or even collegial mutual tolerance, the members of this uncommunal community compete ferociously and engage in notoriously vitriolic polemics with each other. Although modern science has been called the locomotive of all modernity, the scientific community more closely resembles a medieval guild in its hierarchies and career stages of graduate student apprentices, itinerant postdoc journeymen, and PI masters in charge of their own workshops. The reward system is more archaic still, based on mutual recognition by peers, just as aristocratic codes of honor regulated who was qualified to provide satisfaction when challenged to a duel. Nothing about the scientific community we so constantly and casually refer to today is self-evident—least of all its very existence.

And yet as recent events have shown, in the face of two global crises, disastrous climate change, and a deadly pandemic, the scientific community has shown itself capable of consensus and concerted action that even the most cohesive nation-state might envy, much less fractious international organizations like the United Nations or the G-8. Insofar as effective international governance exists, the scientific community is exhibit A. How did it come into existence, and why does it work, despite all of the countervailing forces that have always threatened to tear it apart, from industrial secrecy to national rivalries to plain old personal vanity and greed? What exactly is the scientific community, and how did it so improbably come to be imagined *as* a community?

This book is an attempt to answer these questions in the form of a brief historical overview, from the late seventeenth to the early twenty-first centuries. The disadvantages of combining brevity with a three-centuries-plus survey are obvious: coverage will be of necessity selective and episodic. A fat volume might be (and sometimes has been) written on almost every subsection of every chapter. I hope the references at least will satisfy readers in search of more in-depth treatments. The advantages of this format are threefold: first, a compact account accessible to non-specialists (including scientists) with neither the time nor the patience for a whole shelf full of monographs;

second, a bird's-eye view that highlights the key moments of historical change; and third, insight into how much *has* changed in how scientists have imagined and administered themselves as a collective greater than the sum of its individual members. Quite aside from the value of understanding how current arrangements came about, the recognition that these are neither inevitable nor essential to science can be liberating: thinking about the governance of science as a work in progress can open the door to further progress.

"Governance" is a word of relatively recent currency, with a steep rise in usage after about 1985. It's usually associated with corporations, nation-states, and intergovernmental organizations like the World Bank. To apply the word to science, especially the science of past centuries, stretches its present meaning, so let me clarify from the outset how I will use this term. Like any other organized human activity that exists to achieve certain goals, the governance of science combines an internal ethos (nowadays most enduringly instilled during doctoral and postdoctoral studies) with external incentives (e.g., academic promotion) and sanctions (e.g., forced retraction of flawed or fraudulent journal articles). Like other professions, such as law and medicine, science jealously defends its autonomy against interference from public opinion, economic pressures, and political agendas—ever more difficult in modern democracies where most research is publicly funded in the name of public welfare.

But unlike corporations, governments, or the traditional professions, the governance of science is notably informal (at times even chaotic) and consensual. The institutions that control their incentives and enforce their sanctions, such as universities, journals, funders, and disciplinary societies, are decentralized and only loosely coordinated, if at all. Different institutions—for example, tenure committees and journal editors—can and do tug in different directions, as do the commercial firms and governments that also employ scientists. The ways in which governments paying for research intervene also vary widely, from US senator William Proxmire's Golden Fleece Award skewering the most ludicrous research financed at taxpayer expense to the United Kingdom's emphasis on demonstrable societal impact to the South African government's use of cash bonuses keyed to the number of articles published in international journals. No official mission statements, intergovernmental treaties, boards of governors, or certified regulations rule world science.

In the absence of such visible external mechanisms, the burden of forging and sustaining shared goals and standards shifts to an even murkier internal ethos, which is taught more by example and moralizing anecdotes of good conduct rewarded and bad conduct punished than by codified rules. What counts as good or bad science differs from discipline to discipline or among research groups in the same discipline. Only recently, and usually in the aftermath of scandal, have some fields attempted to articulate and devise ways of enforcing the values of what has now come to be called "research integrity."* Almost none of the current values that constitute this ethos, such as originality,

* One of the most notorious recent examples was the replication crisis in social psychology, which triggered the creation of watchdog organizations such as the UK Reproducibility Network, available at The UK Reproducibility Network (website), accessed December 30, 2022, https://www.ukrn.org/, or the Meta-Research Center at the Dutch University of Tilburg, available at Meta-Research Center (website), accessed December 30, 2022, https://metaresearch.nl/.

open publication, or replicability of results, was always a recognized norm, and new norms—for example, sharing data in public archives—are still being debated. The ethos of science has always been and remains a ragged patchwork, one that has been stitched and restitched many times over several centuries.

Yet somehow what looks to an outside observer (and some insiders) like a free-for-all holds together and gets things done. Still more impressively, consensus gradually crystallizes out of ferocious controversy. Much has been written by philosophers, historians, sociologists, and scientists themselves about the shared standards of acceptable evidence and conduct that make this minor miracle of order out of chaos possible. The question this book addresses is a different one: not what those standards are or how they evolved but how they came to be *shared*. The precondition for governance, whether by institutions or ethos or both, is a collective whose members acknowledge its existence and the legitimacy of its claims upon them. Subjects of a kingdom, citizens of a nation, even the dispersed followers of a religion incarnate their collectives in territory, grandiose architecture, solemn ceremonies, cuisines and holidays, traditions and histories, laws and customs, a common tongue. The scientific collective has none of these props. Nor does it have a monarch, a president, a pope, or a parliament. More than any other imagined community, the work of the imagination that called the scientific community and its predecessors into existence is mighty.

This book is about that work of the imagination. Imagining intellectual communities has a long lineage, as old as the scribal dynasties of ancient Mesopotamia, the philosophical schools of Greco-Roman antiquity, the mandarins of imperial China, or the Vedic scholars of medieval India. But although all of these communities wielded power and influence in the societies they served, their primary declared purpose was the transmission of knowledge, whether of astrometeorological observations, philosophical dialogues, or sacred texts, all deepened and broadened by centuries of commentary and criticism. This essential function of intellectual communities has never been lost: NASA's *Five Millennium Canon of Lunar and Solar Eclipses* includes ancient Babylonian observations, and every year sees a rash of new publications reflecting upon the works of Plato and Confucius. But starting in seventeenth-century Europe, self-consciously new models of intellectual sociability dedicated themselves to coordinated, collective empirical inquiry.

There was nothing new about empirical inquiry, practiced everywhere and always by all human cultures and carefully compiled in Latin herbals on the medicinal properties of plants or in Sanskrit *shastras* on how to rule kingdoms or keep horses or in farmers' proverbs on the signs that portended good or bad harvests. What *was* novel was the aim to vastly expand the scope of both inquiry and compilation and to coordinate these efforts in the present. Instead of a diachronic collective imagined as spanning generations, reaching deep into the past and far into the future, the early modern Republic of Letters, at least its scientific province, additionally imagined itself as a synchronic collective scattered across countries and, as European commercial and imperial ambitions expanded, increasingly across continents as well.

But how to coordinate these dispersed inquirers—their objects of study, their instruments and methods, their divergent theories and conclusions, their assertive and irritable egos? How to motivate them to take notice of and respond to each other, to share their hard-won results, to accept a communal verdict of praise or blame—even to imagine themselves as members of a community with neither schools, nor libraries, nor masters? This was and remains the challenge of imagining the scientific community, and it is where the book begins—and ends.

The three chapters focus on key points of historical inflection in the long-term formation of what we now call the scientific community: first, the halting and for the most part abortive attempts to forge international collaborations among some European savants during the Enlightenment; second, the emergence of international scientific congresses that enacted enduring collaborations and regulations in the late nineteenth century; and third, the emergence of new forms of scientific internationalism and governance in the wake of two world wars and the Cold War of the twentieth century. Each of these episodes played out in political, economic, military, and cultural contexts that set the stage: for example, the efflorescence of international scientific congresses circa 1900 would have been unimaginable without worldwide steamship and telegraph networks, just as Enlightenment scientific expeditions would have been impossible without colonial outposts in the Americas, Asia, and Africa, as would also have been the International Geophysical Year in 1957–58 absent Cold War politics. But none of these enabling factors could have been mobilized to scientific ends without impulses from a few influential individuals who galvanized their colleagues and sometimes their governments into action.

As an intellectual historian, I attend closely to how these prime movers conceptualized and justified such collective efforts, which were always expensive and time-consuming, often at odds with individual interests and local research agendas, and sometimes fraught with danger. Historians such as Benedict Anderson and Linda Colley have taught us to think of modern nation-states as imagined communities, a resonant phrase in which the word order parallels the temporal sequence: before a community can come into being, it must first be imagined, and that vision of what could be must be vivid and compelling enough for people to will it into existence. Just as importantly, once realized, the vision must be sustained and anchored in loyalties firm enough to resist the centrifugal forces that tear communities apart. The work of the imagination is mighty in the history of every new kind of community, and the scientific community is no exception. Indeed, the scientific community strained the imagination beyond anything ever attempted by communities that occupied territories, shared a language or religion, and were reinforced by daily personal encounters.

The imagination is also fertile in tangible consequences. Each of the three episodes narrates how imagining the scientific collective as a Republic of Letters, a world project, or *the* (emphatically in the singular) scientific community not only changed science but also individual lives and occasionally the world. For example, without the new models of intergovernmental scientific collaboration pioneered by the World Meteorological Organization after World War II, the global observation networks that delivered the mounting evidence of planetary climate change would also probably never have been constructed. More fundamentally, sustained scientific coordination and collaboration are always premised on the individual's will to participate in such a collective, almost always at the price of at least short-term losses in freedom to pursue one's own research priorities in one's own way. The scientific polity is no different from any other polity in this regard and poses the same quandaries of costs and benefits to its members. But as historians have shown and political theorists have acknowledged, no viable collective is ever just the result of cool cost-benefit calculations. It is a shared vision of what it would *mean* to be part of a collective that surmounts hesitation and commands allegiance. What that collective vision should be for science already has a 350-odd-year history, and the work of the imagination is still ongoing.

The Republic of Letters

Pen-Pal Science

1.1. Introduction: The End of Splendid Solitude

On a cold November night in 1619, French soldier of fortune René Descartes (1596–1650) dreamt that he would single-handedly revolutionize all of philosophy, which then included mathematics and the study of nature. He set about deducing everything from the orbits of the planets to the nature of fire to the beating of the heart from a few first principles, or "laws," as he would later call them, on the model of the theorems deduced from axioms and postulates in Euclidean geometry. But by 1637, when he published the first fruits of his solitary inquiry prefaced with a discourse on method, he had to admit defeat. Although he had achieved remarkable things—inventing analytic geometry, explaining the rainbow, deriving the angles of reflection and refraction of light—he acknowledged that his methods could not be stretched to discover the causes of the myriad particulars of the observable world. Instead, experiments would be needed, lots of them. Descartes still clung to the hope that he could accomplish even this enlarged program of empirical research on his own and urged his readers to send him money for equipment and paid assistants rather than to try to perform the experiments themselves. Quite apart from the difficulty of getting volunteers to do as they were told, "they would inevitably wish to be rewarded by having certain difficulties explained to them, or at any rate by compliments and useless conversation," all a great waste of his precious time. Nor did he think he could benefit from the observations and experiments of others (probably poorly explained or downright wrong, in Descartes's opinion), much less from engaging in debate about his novel ideas (he'd already foreseen all objections and refuted them).

Descartes was probably the last major thinker to believe that science could be conducted in splendid solitude. In the decades after his death in 1650, academies dedicated to the collective investigation of those dazzlingly various and variable natural particulars that had stymied Descartes's deductions sprang up all over Europe: the Academia Naturae Curiosorum of the Holy Roman Empire ("Academy of Those Curious about Nature," established 1652, later known as the Leopoldina), the Accademia del Cimento in Florence ("Academy of Experiment," established 1657), the Royal Society of London (established 1660), the Paris Académie royale des sciences (Royal Academy of Sciences, established 1666). All enlisted their members and correspondents in the endless labor of empirical inquiry and saw their own role as one of collecting and coordinating the results.

Their annals are crammed with observations and experiments on heat and cold, blood transfusions from a sheep to a human, the nature of light and color, shanks of meat that glowed in the dark, daily weather observations from near and far, exotic flora and fauna sent in by travelers and missionaries, new stars in the heavens, and monstrous births. New instruments such as the vacuum pump, the barometer, the telescope, and the microscope multiplied the objects of inquiry: Would smoke rise in a vacuum? What were those strange, earlike protrusions around the planet Saturn? How did air pressure decrease as altitude increased? They could also turn the most mundane things into marvels: a common kitchen-variety fly viewed under the microscope became a tiny warrior sheathed in shimmering blue armor. Dazzled by the blooming, buzzing confusion of things to observe, the academicians sought reinforcements in a legion of correspondents, ranging from provincial doctors to sea captains to Jesuit missionaries, who sent in their observations of anatomical rarities, winds on the high seas, and the stars of the southern hemisphere. Whereas earlier generations of naturalists had carefully preserved and recycled observations from Pliny the Elder's (c. 23–79 CE) *Natural History* or Claudius Ptolemy's (c. 100–170 CE) astronomical treatises as scant resources, their successors in the seventeenth century, eager consumers of the curiosities brought back from voyages to the Far East and Far West and the books pouring from the printing presses, nearly drowned in the first data deluge.

In this all-hands-on-deck situation, the model of both inquiry and the inquirers was transformed. Natural inquiry was no longer about certain knowledge of universal causes, the ideal of medieval natural philosophy still reverberating in Descartes's mammoth project. Instead, it was first and foremost an inventory of natural particulars, to be eventually and then only probabilistically explained by as-yet-undiscovered causes. The inquirers were no longer lone scholars ensconced in their studies or lecturing a roomful of students from a canonical text but rather members of a collective that occasionally met in person, as at meetings of the scientific academies, but more often remotely, in letters exchanged among individuals who would perhaps never meet face-to-face and who might share neither mother tongue nor religion nor social rank nor education—nothing except their membership in that vast, vague polity known as the Republic of Letters.

This chapter is about the scientific collective as imagined within the Republic of Letters. It was by no means the first learned collective, but it was the first to try to organize itself to work together in the present, as well as to transmit the knowledge of the past, and moreover to envision the future as progress rather than decline. It was cosmopolitan but not international: nations were more hindrance than help to the Republic of Letters, their wars and rivalries wreaking havoc with the post, the book trade, and scientific expeditions. As their names suggest, scientific academies such as the Royal Society of London were metropolitan rather than national institutions; the Paris Royal Academy of Sciences lumped anyone not resident in Paris, whether Frenchman or foreigner, into the category of "étrangers." On the very rare occasions when nations did collaborate in the name of science, it was the personal interest and vanity of their princes rather than any spirit of internationalism that prompted them to do so.

The operative word in the Republic of Letters was "republic." Insisting upon self-rule, its members vehemently opposed the imposition of social hierarchies upon their own hierarchy of intellectual merit—and defended their right and their right alone to judge merit even more vehemently. They also embraced the cardinal virtue of early modern republicanism, namely, freedom—which they interpreted as freedom to criticize each other with a ferocity that sometimes threatened to tear apart the Republic of Letters and its most prominent institutions, the academies. All in all, early modern republicanism was not a promising conceptual framework within which to transform a swarm of solitary, dispersed, and fiercely independent scholars into a working collective. The later image of science as an arms-linked international community was completely at odds with the rival vision of the Republic of Letters as a state of nature in which each individual waged war on everyone else with sharpened quill pen and still sharper tongue. How to imagine a collective in this Hobbesian state of savage freedom without ending up with Hobbes's absolutist Leviathan?

1.2. The Dance of Distance and Proximity

What might such an unprecedented collective look like? Various experiments were tried; most failed as dismally as Descartes's had. Two images of intellectual community, both European, and both circa 1700, give some idea of the range of possibilities. The first shows two periwigged men perusing volumes of portraits of scholars from many disciplines, times, and places. Behind them bookshelves of many more such volumes, arranged by discipline, fill the room; the volumes they are consulting rest upon a cabinet containing medallions struck with yet more likenesses of savants both past and present (figure 1). The collectors have carefully annotated each engraved portrait with the birth and death dates of its subject, as well as details about whatever is known of his (and the subjects are overwhelmingly but not unexceptionally male) life and works. In contemplating these portraits, the two men actually present are committing the whole history of science and letters to memory, fortifying their recollection of "the most remarkable circumstances of the lives and writings" of absent scholars by gazing on their portraits. This is an image of intellectual community at a distance, in which luminaries remote in both time and space are virtually gathered together. Distance had its advantages. As the nineteenth-century British historian Thomas Babington Macaulay explained about why he preferred communing with great minds at a distance in his library, these ghostly companions were always on their best behavior: "With the dead there is no rivalry. In the dead there is no change. Plato is never sullen. Cervantes is never petulant. Demosthenes never comes unseasonably. Dante never stays too long."

The second image shows members of the Paris Royal Academy of Sciences gathered around a table where they are dissecting a duck together (figure 2). In addition to the two figures in shirtsleeves and aprons actually dismembering the duck, academicians observe the proceedings, point to passages in anatomical treatises, discuss among themselves, and inspect the skeletons of the other animals hung on the wall for the purposes of comparison; an artist in vaguely classical robes sketches the duck's innards. This is an early experiment in collective inquiry and authorship: in one of the fledgling Academy's first projects, its members aimed to produce a comparative anatomy of animals as a group, not as individual authors. At the Academy's quarters in the royal library in the rue Vivienne, a team of anatomists, assisted by several artists, dissected a camel, a bear, a pelican, an ostrich, and other such animals as came their way, usually via a death at the royal menagerie. Their short-lived ideal of group authorship was not

Icones
Separatim
Collectae
Icones
Theologorum
Icones
Reformat.
Icones.
Pontif. etc.
BILDER
CABINET
Ico: Iuris
Icones.
Medicorum
Ico: Philo
MUNZ – CABINET

Figure 1

the modern *Science* or *Nature* article signed by a dozen or more names of individuals but rather the melding of all individuality into a single corporate identity: "la compagnie," a word redolent of associations with guilds and other early modern corporations. The collaboration was premised on a strict egalitarianism within a small, in-person elite: as the preface announced, all observations were vouchsafed by academicians who had "eyes to see such things." By submerging authorial vanity in the "compagnie," the members of the Company would not be tempted to exaggerate the novelties of their discoveries nor to sacrifice an inconvenient fact to "fine reasoning or a new system"—at least that was the theory.*

The reality in both cases was a good deal less chummy. In the case of the duck-dissecting academicians, the idea of collective authorship and inquiry was soon abandoned; worse,

Figure 2

* The model proved unsustainable. Already in the second, expanded edition Perrault's authorship was flagged in a subtitle: *Mémoires pour servir à l'histoire naturelle des animaux; Dressez par M. Perrault* (Paris: Imprimerie royale, 1676).

rivalries among members became so venomous that the academy's revised regulations of 1699 had to explicitly forbid name-calling at meetings and in print. In the case of the savants communing in their studies with the visages of dead and distant colleagues, would-be collectors had to be admonished not to degrade the ranks of the learned by specializing in portraits of scholars who had been the sons of prostitutes or churlishly vain and arrogant. Animosity, not amiability, set the tone for the Republic of Letters.

Dead or alive, thousands of miles distant or at one's elbow, colleagues were also potential rivals for the fame and honor that was always by definition in short supply. Then as now, in this symbolic economy of the limited good, praise for one person was perceived as blame for another, and vice versa: your Nobel Prize is someone else's missed Nobel Prize. In this regard, the only difference between the Enlightenment Republic of Letters and today's scientific community is that the savants of the seventeenth and eighteenth century were considerably less inhibited in venting their spleen toward each other in public. The French philosopher Pierre Bayle (1647–1706), editor of the first book review journal, *Nouvelles de la République des Lettres* (News from the Republic of Letters, founded in 1684), and himself an acid-penned polemicist, described the Republic of Letters as a bellicose state of nature. In the name of truth, "War was innocently waged against one and all: friends must be on guard against friends, fathers against their children, fathers-in-law against sons-in-law; it is like the age of iron."

And yet these warring savants needed each other, and they knew it. First, by the second half of the seventeenth century it had become clear that no one person, no matter how brilliant or diligent, could advance science and scholarship alone. There was so much to observe, so many experiments to perform, so many places to explore, that no individual, no matter how industrious and ingenious, could hope to make even a start in a single lifetime. All of the early scientific academies recognized the imperative to join forces into some kind of collective, and a trans-generational, trans-continental collective at that. Francis Bacon's (1561–1626) fragmentary utopia *The New Atlantis* (1627) supplied one frequently invoked blueprint for such a collective: the House of Salomon, a vaguely monastic order that collected observations and performed experiments to gain "knowledge of Causes, and secret motions of things, and the enlarging of the bounds of Human Empire, to the effecting of all things possible." But no one, including Bacon himself, though he served as Lord Chancellor to King James I of England, succeeded in finding the necessary funds and manpower to create such an institution, much less to impose the authority of the abbot-like "Father" of the House of Salomon upon individual researchers.

Yet nature's scale dwarfed all individual human efforts. As Johannes Laurentius Bausch (1605–1665), first president of the Academia Naturae Curiosorum, explained in that academy's founding document, the works of God in the vegetable, mineral, and animal realms were "innumerable, and of such an extent, that the lifetime of a single person does not suffice to investigate and know them precisely no matter how ardent the desire for knowledge; [but] this failing, which lies in the fact that the lifespan of an individual is too short to do justice to the innumerable natural phenomena to be researched, can perhaps be compensated for by several people banding together to work with shared dedication."

Second, the citizens of the Republic of Letters trusted no one but themselves to confer the fame and glory they so craved. Several of the early scientific academies, such as the Royal Society of London and the Paris Royal Academy of Sciences, had inducted some members on the basis of their social rank rather than their scientific achievements, a sign of how precarious these new institutions were and how parasitic scientific honor was upon aristocratic honor. But by the early decades of the eighteenth century, scientific self-confidence had risen to the point that the patronage of lords and ladies, much less reliance on their judgment, could be proudly rebuffed. It was not that the champions of the Republic of Letters rejected aristocracy in principle; they simply erected a supplementary aristocracy of merit alongside one of inherited rank. In both cases, only peers could confer recognition upon peers—a word that was itself redolent of nobility. But in contrast to hereditary titles, true scientific merit eluded easy recognition, especially by envious rivals more eager to blame than to praise. The French mathematician and *philosophe* Jean d'Alembert (1717–1783) traveled in both literary and scientific circles and judged the latter to be the more vindictive: "A bad epigram is sometimes all the vengeance a poet takes, but that of a savant is more constant and premeditated." Yet to trust the weather vane and inexpert opinion of the general public would compromise the proud autonomy of the Republic of Letters, especially its scientific province. The solution? D'Alembert spoke for many in vaunting the virtues of distance: only those savants remote in both time (posterity) and space (foreigners) could muster the necessary impartiality to judge merit fairly. The glory bestowed by posterity might be more lasting, but that of foreigners had the distinct advantage of awarding laurels while the laureate was still around to rest upon them. Hence d'Alembert's ambivalence about becoming too close to foreign colleagues: "The closer one becomes to foreigners, the more they lose that character of posterity for which the distance of space is at least necessary, in default of the distance of time."

Every intellectual community is a dance between proximity and distance. Intellectual communities have been continuously invented and reinvented from the scribal schools of ancient Mesopotamia to the listeners of weekly podcasts that discuss the latest preprint in virology. Many of these communities have centered on the transmission of knowledge from one generation to the next, from the schools that prepared aspirants for the imperial examinations in Song Dynasty China to modern universities all over the world. Others were centered on agonistic displays of wit and learning in the marketplace or princely court, as in the story of St. Catherine of Alexandria winning a debate against fifty pagan philosophers before the emperor, or when two Renaissance mathematicians dueled over solutions to cubic equations in a Milanese square, or Indian pandits debated philosophical subtleties for the delectation of a medieval prince. All of these communities combined proximate, in-person relationships with more distant ones: teacher and student faced each other in the classroom, but the texts on which the teacher lectured were often centuries old, embellished by layer after layer of commentary as each generation of scholars revived the ideas of past thinkers for the present audience. Similarly, scholars and scientists today meet each other in person at universities and conferences but also in the pages of books and journals and online.

There was nonetheless a novel element to the collectives invented by the Republic of Letters. In contrast to those established to assure the continuity of knowledge of intellectual traditions by copying, teaching, elucidating, criticizing, and accumulating texts and techniques over centuries and sometimes millennia, the new-style intellectual collectives of the Enlightenment Republic of Letters aimed to create collaborations in the here and now. Past intellectual collectives had often been imagined as family genealogies, such as the philosophical schools descended from a founding master (for example, the Epicureans or the Confucians) or as ersatz lineages of teachers and students—of which the term "Doktor-Vater" is a remnant. The Republic of Letters instead imagined itself as a polity, geographically dispersed, divided by nationality and confession, riven by rivalries, but united by shared objects of curiosity, consensus concerning status and standards, and crisscrossing networks of correspondence. The Republic of Letters represented itself as a map, not a family tree.

The Republic of Letters was arguably the first intellectual community to confront the challenge that still defines modern science and scholarship: how to balance competition and cooperation, hierarchy and equality, individuality and collectivity, proximity and distance in a quest for knowledge that could no longer be pursued alone. Let us return to those two images of the life of the mind pursued close up and far away. One of these imagined communities held its members at a distance; the other gathered at least some of them in the same room. But both sought to meet a novel challenge in the world of learning: how to pursue knowledge collectively *at the same time* with fellow inquirers near and far. Recall the image of the family tree:

intellectual traditions were by no means static, but their primary goal was handing down a legacy *across time*, just as in a family. Successive generations and exchanges among traditions enlarged this inheritance with new ideas and discoveries; contemporaries honed their arguments against each other in treatises and disputations. But they rarely collaborated or even coordinated with each other. The collective efforts of the intellectual community were directed toward assuring the continuity of past and future, not toward coordinating inquiries in the present. Now recall the image of the map: this intellectual community must overcome not only distance in time but also in space. Its members, both the scholars and the scientists, still drew upon past archives, just as scholars and scientists do today: ancient Chinese observations of supernovae are as important to astronomers as ancient Greek tragedies are to literary scholars. But the intellectual community of the map must somehow also knit its dispersed members into a network woven of thick cross-hatchings of correspondence, exchange of specimens and publications, and even occasionally of concerted collaborations.

By 1750, both of the imagined collectives with which this section began had gone as extinct as Descartes's grandiose effort at solo science. Savants were neither communing with portraits of the distinguished dead nor were they trying to mind-meld with the living. Instead, they were mostly publishing as individuals, working in the private spaces of their own homes—and conducting voluminous correspondence with colleagues near and far. This became the primary model for collective science in the Enlightenment Republic of Letters. Here is one such typical citizen of the mid-eighteenth-century Republic of Letters. This image shows the celebrated Genevan naturalist Charles Bonnet (1720–1793), famed for his studies of plants and insects, alone in his study except for an amanuensis to whom he is dictating letter after letter containing reports of his latest observations and experiments (figure 3). Although Bonnet was a corresponding member of several scientific academies and maintained a brisk exchange of letters with other prominent naturalists, the closest he ever came to an international collaboration was to suggest to one of his correspondents, the Italian naturalist Lazzaro Spallanzani (1729–1799), that he might repeat some of Bonnet's earlier experiments. International science during the heyday of the Republic of Letters was as a rule pen-pal science.

Figure 3

There were two notable eighteenth-century exceptions to this rule: the international expeditions mounted in 1761 and 1769 to observe the Transits of Venus; and the meteorological observing network created by the Mannheim Societas

 Meteorologica Palatina (Mannheim Meteorological Society) from 1780–1792. Although both efforts were hailed at the time and thereafter as pioneering examples of international collaboration, they were both short-lived and their scientific contributions were meager. Both depended crucially on the heroic engagement of a few individuals and never gained a firm institutional foothold, despite the fact that each was sponsored by established institutions of the time: in the case of the Transits of Venus expedition, several academies and royal courts; and in the case of the meteorological network, the court of Karl Theodor, Elector of the Palatine (1724–1799). Both episodes are rich in lessons in how the Republic of Letters failed to become the scientific community.

1.3. Mobilizing the Academies: The Transits of Venus

In 1716, the English astronomer Edmond Halley (1656–1742) published an article on how terrestrial observations of the transit of Venus across the sun's disc might be used to determine the distance of the earth from the sun and the scale of the entire solar system. Transits of Venus were rare events, occurring about once a century in pairs separated by an interval of approximately eight years. Since the next pair of transits was predicted for 1761 and 1769, astronomers would have plenty of time to prepare. Halley thought all that would be needed would be a telescope, a reasonably good clock to time the transit, and a certain amount of care and diligence on the part of the observer. But there was a catch: like solar eclipses, the transits would only be visible from certain parts of the globe, and observing stations would also have to be sufficiently far apart to guarantee a long enough baseline for precise measurements of the angle of solar parallax. Halley suggested Bencoolen, an outpost of the East India company in Sumatra, or Batavia (now Jakarta, Indonesia), a trading center of the Dutch East India company, as possible observing stations, and there the matter rested until the spring of 1760.

On April 27, 1760, the French astronomer Joseph Nicolas Delisle (1688–1768) sent out a memoir containing both observing instructions and a world map of regions of visibility of the upcoming transits of Venus to some two hundred addressees all over Europe and beyond, including both the Paris Royal Academy of Sciences and the Royal Society of London (figure 4). Delisle was doubly qualified to launch an international observing initiative of the sort Halley had envisioned. First, he had already attempted to mobilize the support of academies in 1752 and 1753 for the observation of the transits of Mercury, and succeeded in getting Paris and Stockholm at least to coordinate their work, with further participation by St. Petersburg, Berlin, Bologna, Montpellier, and London. Second, and even more important, Delisle was the go-between *par excellence* of European Enlightenment science, a man who traveled everywhere, who knew everyone who was anyone, and whose correspondence network had a wider circulation than most academic proceedings and was just as densely packed with the latest scientific news. He had met Halley in London in 1724, spent decades at the St. Petersburg Academy in Russia, kept up contacts with influential members of academies in Paris, London, Berlin, and Stockholm as well as with Jesuit missionaries in Peking, Pondicherry, and Quebec, and even cultivated the widows of astronomers whose observations he hoped to purchase. The list of recipients of his world map for the 1761 transit of

 Venus reads like a Who's Who of Enlightenment science: Leonhard Euler in Berlin, Daniel Bernoulli in Basel, Francesco Zanotti in Bologna, Pehr Wargentin in Stockholm, plus the editors of various journals and monarchs such as the empress of Russia. It is unlikely that any of the academies would have organized an observing expedition without Delisle's prodding, and still more unlikely that they would have made any attempt to coordinate their efforts.

As it was, coordination was more a matter of principle than of practice, and competition proved at least as effective a goad to participation as any spirit of international scientific cooperation. Let's return to Delisle's April letter to the Royal Society of London. It did not reach the Royal Society of London until June, delayed by the Seven Years' War (1756–63) then raging between Britain and France. Although the Royal Society agreed to send observers to St. Helena in the south Atlantic Ocean and Bencoolen in Java, both under the control of the East India Company, the only way to win funding for the expedition was to appeal to national rivalries, not to international scientific solidarity. When Lord Macclesfield in his capacity as president of the Royal Society wrote to the Duke of Newcastle to request royal funding, he pointed out that "the French King is sending observers not only to Pondicherry and the Cape of Good Hope, but also to the Northern Parts of Siberia; and the Court of Russia are doing the same to the most Eastern Confines of the Greater Tartary"; it was therefore a matter of "the honour of His Majesty and of the Nation in general" to bankroll the British expedition. The ploy worked: King George III duly approved £800 for the expedition to St. Helena.

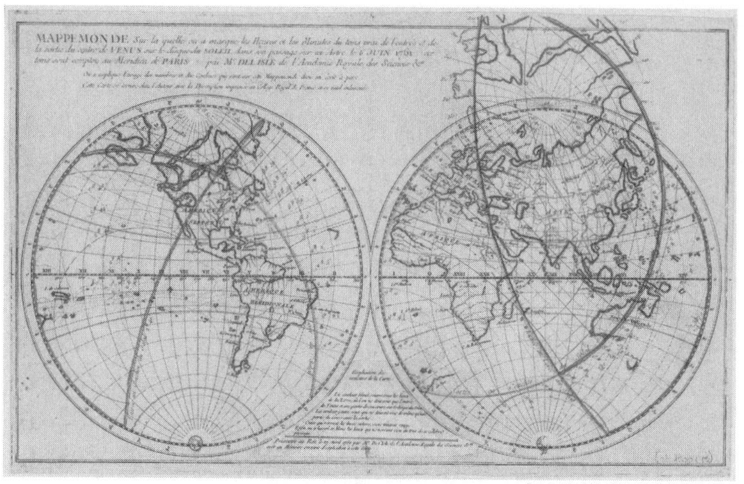

Figure 4

But national rivalries did not always work to promote the cause of science: although the British admiralty granted the French astronomers a letter of safe passage to the Isle of Rodrigues in the Indian Ocean, it was summarily ignored by the British man-of-war that attacked and sacked the French possession in late June 1761. For their part, French forces had meanwhile captured Bencoolen, one of the British observing stations, obliging the emissaries of the Royal Society, Charles Mason and Jeremiah Dixon, to decamp to the Cape of Good Hope instead. Spain refused to grant the British permission to observe the transit of 1769 in California, then a Spanish territory, favoring their French allies instead. Despite protestations of neutrality, the astronomers were also caught up in the hostilities of the

Seven Years' War, which extended well beyond Europe's borders to colonial and commercial settlements in Africa, Asia, and North America.

Both the 1761 and 1769 observing expeditions reported adventures and misadventures aplenty. The French astronomers sent to Rodrigues had their possessions and instruments repeatedly ransacked by British seamen; another French astronomer finally made it to Pondicherry in India after a perilous voyage and eight years of waiting, only for clouds to obscure the sun on the day of the transit; the Siberian expedition arrived so late in the season that they feared the ice on the frozen rivers would no longer bear the weight of their heavily laden sleds; the observing party sent from Massachusetts to Newfoundland complained of the venomous insects that plagued them day and night; some observers perished. Everyone had to deal with delicate instruments damaged en route, and everyone watched the weather on the day of the transit in a fever of anxiety lest the sky cloud over at the last minute. Jean-Baptiste Chappe d'Auteroche (1728–1769), the French observer who had journeyed from Paris to Tobolsk in Siberia with the support of the St. Petersburg Academy, wrote on the day of the transit in 1761, "I was seized with an universal shivering, and was obliged to collect all my thoughts, in order not to miss it." But he steadied himself at the last minute with the thought that posterity, those distant and impartial judges, would find his observations useful "when I had quitted this life."

In the end, over 120 observers at 62 stations, from Peking to Newfoundland, from Tobolsk in Siberia to the Cape of Good Hope in southern Africa, attempted to measure the angle of solar parallax during the first transit in 1761. Even more participated in the observations of the second transit of 1769, 151 observers at 77 stations, from Hudson Bay to Tahiti. Their efforts were only loosely coordinated, if at all: no attempt was made to standardize instruments; observers included novices as well as seasoned experts; observing conditions varied from excellent to abysmal. Unfortunately, but unsurprisingly, the all-important values for the angle of solar parallax diverged so widely that no reliable conclusions could be drawn about the absolute value of the astronomical unit (the distance from the earth to the sun)—the determination of which had been the aim of the entire arduous, expensive undertaking.* Astronomers set to squabbling among themselves as to whose values were most trustworthy. The Republic of Letters once again fractured into warring factions, its habitual state.

None of this augured well for sustained international scientific collaborations—nor had any of the moving spirits behind the expeditions, not even the well-connected Delisle, ever expected that this twice-in-a-century event would yield anything of the kind. It had been possible to mobilize international support in part because it was a one-shot (or rather two-shot) deal: the transits of Venus would not occur again until 1874 and 1882. It is questionable whether the effort can even be called international, in the sense of involving several nation-states. As the institutional representatives of the Republic of Letters in their respective countries, the academies in Paris, London, Stockholm, and St. Petersburg had managed to mount observing expeditions after receiving Delisle's call. But the actual patrons of the undertaking were rulers such as George III of Britain,

* Values in 1761 ranged from 8.5" to 10.5", an enormous margin of error.

Catherine II of Russia, or Christian VII of Denmark, who took a personal interest in the expeditions and assessed their generous support in terms of the prestige accrued in their competition with other royal courts. Mutatis mutandis, the same went for the many Jesuit observers, from Vienna to Madras, whose participation redounded to the greater glory of their order. When the exhausted but exhilarated French observers on the Isle of Rodrigues raised a glass in celebration of their hard-won observations on June 6, 1761, they toasted first the king of France and then all of the other astronomers who had succeeded in observing the transit that day—an order of precedence that accurately reflected the loyalties implicit in the very structure of the expeditions. Observing the transits of Venus on a global scale for the first time was an impressive, even heroic feat. But the undertaking was neither a true collaboration nor truly international—and it was never meant to outlast the transits themselves.

Nonetheless, the expeditions lingered much longer in scientific memory, both their successes and their failures. The next notable attempt to create an international scientific collaboration—this time, a more sustained one—would learn lessons from both the 1761 and 1769 expeditions about how to choreograph the dance of distance and proximity among the citizens of the Republic of Letters.

1.4. Centralizing Everything: The Mannheim Meteorological Network

In 1771, the mathematician and astronomer Johann Heinrich Lambert (1728–1777), member of the Berlin Academy of Sciences, read a paper to that body on how to discover the "general laws" of the weather worldwide. That there were such laws, Lambert had little doubt. For decades, weather watchers had noted patterns—for example, that the weather in Zurich, Switzerland, reached Upminster, England, about five days later. But to date, weather observations had been too spotty and sporadic to reveal the sturdy regularities Lambert was certain existed. What was needed was long-term, systematic, coordinated barometer and thermometer observations on a global scale. In a map reminiscent of Delisle's mappemonde for the transits of Venus, Lambert divided the globe up into twenty triangles, with thirty-two observing stations positioned at the centers and points of intersections of the triangles. Lambert admitted that the project might be expensive but quite modest compared to the "considerable sums [spent] on a few astronomical observations"—an obvious reference to the transits of Venus expeditions—and of infinitely greater utility to everyone, from farmers to mariners. And if the price tag was too high and the organizational effort too great, the number of observing stations might be located in "places where the commercial nations have established colonies or where there are missionaries"—another reference to the strategy followed by the transits expeditions in their choice of observing posts beyond Europe. So closely intertwined were such global scientific ventures with global commercial imperialism that Lambert suggested that the Royal Society of London should take the lead in recruiting the participation of further academies "established in the other commercial countries." Just as Lambert hoped that such a grand observational network might discover meteorological laws that could compete with those of astronomy in accuracy and reliability, so he took the transits of Venus expeditions as his template for global scientific cooperation.

Lambert's plan for a dispersed network of weather watchers was not novel. Since the very earliest scientific academies in the seventeenth century, repeated attempts had been made to establish such networks, starting with the Accademia del Cimento in Florence (Academy of Experiment, 1657–1667). The archives of both the Royal Society of London and Paris Royal Academy of Sciences contain many tables of barometer and thermometer readings sent in by correspondents at home and abroad. None of these networks survived more than a decade or so, dependent as they were on the wavering dedication of volunteer observers. Nor was the data they produced of much use, for much the same reason: observers took measurements at unstandardized times with uncalibrated instruments or no instruments at all. Volunteers might be country parsons, members of provincial academies, or municipal doctors—the latter especially interested because of a Hippocratic belief in a correlation between the weather and outbreaks of epidemics. All had other duties, and very few met the standards of the ideal observer set down by the French Oratorian Louis Cotte (1740–1815), who administered yet another short-lived network under the auspices of the French Royal Society of Medicine (1774–1793): great exactitude, freedom from preconceived notions, orderliness—and a willingness to be "entirely devoted to this sort of occupation, renouncing for it almost all other business and all pleasures"—not to mention living in the same place for years on end, never traveling, and always being at home for the fixed times of observation. No wonder these volunteer networks petered out, one after another. A survey of all German meteorological observers from 1750 to 1880 showed that only 14 percent were active for twenty years or more—yet that was still too short a timeline to yield the laws Lambert and others sought.

One of the addressees on Delisle's list for the dissemination of his mappemonde in 1760 was neither an academy nor an individual savant: "His Royal Highness the Elector Palatine of Mannheim." Karl Theodor, prince-elector of the Palatine and later Bavaria, was the very model of the enlightened eighteenth-century monarch. Visits from Voltaire and Mozart enlivened his court at Mannheim. He inherited the former's private secretary and commissioned the opera *Idomeneo* from the latter. His modernizing projects still define the cities in which he resided (for example, the Englischer Garten in Munich). A patron of the sciences as well as the arts, he had responded to Delisle's 1760 call for observers by writing to the Royal Society to volunteer the services of his court astronomer, Christian Mayer, FRS, to wherever the Royal Society might choose to send him, all expenses paid by the Elector. Although the Royal Society did not respond (another example of failed coordination), Karl Theodor's ambitions to elevate his court's scientific reputation were undamped. With the help of Johann Jakob Hemmer (1733–1790), the energetic director of the royal cabinet of scientific instruments and secretary of the Mannheim Academy of Sciences, Karl Theodor launched the Mannheim Meteorological Society, the most successful of all Enlightenment scientific collaborations.

Once again following Delisle's playbook for the transits of Venus, the Mannheim Society sent out a circular describing the new observational network to academies all over Europe and beyond, as well as to some individual savants. But unlike the transits mobilization, the Mannheim call for observers contained not only detailed instructions on how and when the meteorological observations should be made and recorded; potential observers were also promised special instruments fabricated and sent at the Elector's expense: a barometer, two thermometers (one to measure the temperature in the sun and the other in the shade); a feather hygrometer; a wind vane; an electrometer; and a vessel for measuring rainfall—all constructed according to the most exacting scientific specifications and affixed with an inscription bearing the Elector's name. Everything about the network would be centralized and standardized, from the instruments to the forms and symbols for recording observations to the daily hours of observation (7:00 a.m., 2:00 p.m., 9:00 p.m., which became known among meteorologists as the "Mannheim hours"). New scales developed by the Mannheim Society for assessing cloud cover and wind force were still in wide use in the late nineteenth century. All of the observations would be collected and published in the *Ephemerides* of the Mannheim Society, again bankrolled by the Elector.

Despite these careful preparations and lavish support, all did not go entirely according to plan. Although the call for observers netted stations in fifty locales, almost all were in Europe; Hemmer's attempts to gain a foothold in Dutch Batavia failed, and he had to settle for Harvard College in Cambridge, Massachusetts, as the outermost post in the network. Once again, the Royal Society failed to reply, an outsized loss because of Britain's overseas possessions. As Lambert had noted, it was the commercial nations with colonies that were best equipped to pursue global science in the age of empire, and even the Elector's deep pockets could not make up for Mannheim's landlocked deficiencies in this regard. (The closest Karl Theodor ever got to acquiring a port city was his failed attempt to exchange Bavaria for the Austrian Netherlands.) Almost all of the instruments arrived at their destinations in shards, and a second (and sometimes a third) shipment had to be dispatched by special courier. Of the fifty original observing stations, only thirty-seven reported regularly, and only eight for the entire duration of the project. Significantly, half of these true-blue observers were located in the Elector's own lands in the Palatinate: centralization and coordination were no substitute for proximity and personal loyalties.

Nor were they a substitute for personal engagement. The project suffered a near-mortal blow when Hemmer died in 1790 and was already in disarray when the French occupation of Mannheim in 1795 delivered the coup de grâce. Hemmer, who had favored academies and universities as observing stations because such institutions would survive the death of an individual observer, had failed to anticipate the fragility of his own institution, the Electorate of the Palatine. The Mannheim Meteorological Society also collapsed after publishing the twelfth volume of the *Ephemerides*. All the weather-watchers of the Republic of Letters understood how precarious their observational networks were, hanging by the thread of an observer's whims and lifetime. Cotte, whom we have already met in the context of the French medical meteorological network and who was also one of the observers in the Mannheim network, knew that he would not live to see the fruits of his own observations, those laws of the weather for which Lambert had yearned. Like the trembling astronomer Chappe, about to observe the transit of Venus in Siberia, Cotte consoled himself with thoughts of a grateful posterity: "A true savant does not limit his work to the present time, he works for the entire society, and as that society does not perish, he extends his perspective beyond his century."

1.5. Conclusion: Where Was Community?

But, as British prime minister Margaret Thatcher once notoriously asked, where was society? The Republic of Letters was more like the state of nature than a state. Its leading institutions, the metropolitan academies of London, Paris, St. Petersburg, and Berlin, could serve as letterboxes to which calls for volunteers could be addressed, but in the end, everything depended of the initiative of a few well-connected individuals and their wealthy patrons to make things happen and keep them going. Without the wide personal networks of Delisle in St. Petersburg, Hemmer in Mannheim, and a few other savants who had visited or been members of multiple academies, it is unlikely that either of the two most impressive eighteenth-century scientific collaborations, the transits of Venus expeditions and the Mannheim meteorological network, would ever have gotten off the ground. Both of them depended crucially on funding from well-disposed monarchs—again, a matter of individually conferred favor rather than institutional backing. Wherever one looks for the sturdy structures associated with society, one finds instead a few bustling go-betweens, endlessly paying visits and writing letters. These self-appointed diplomats of the Republic of Letters cultivated the proximate, personal connections that counterbalanced competition, envy, distrust, and other distancing forces. When these go-betweens died, their globe-spanning networks died with them.

Nor did the prospects for sustained international scientific collaborations improve when late-eighteenth-century revolutions deposed monarchs, although the revolutionary governments were if anything even more eager to bolster their prestige by embracing Enlightenment science. Quite aside from the disruptive wars sparked by the American and French Revolutions, revolutionary patronage of science took on an ideological tinge usually absent from the personal patronage of princes. One last vignette illustrates how these new political configurations could contaminate efforts at international scientific coordination, despite powerful state backing and strong personal bonds.

In 1798, another scientific go-between, the French astronomer Joseph-Jerôme Lalande (1732–1807), announced a visit to another small German court, this time that of Ernest II, Duke of Saxe-Gotha-Altenburg (1745–1804). Lalande had been Delisle's student, assistant, and eventually successor at the Collège de France; he had also helped drum up support for the transits of Venus expeditions and funneled observations from his foreign contacts to Cotte's meteorological network. Like Delisle, he was an indefatigable traveler, making a point of visiting the local academy of every town he passed through, and was himself a member of both the Berlin and Paris academies. He was also a gracious host to visiting scientists in Paris, including the court astronomer at Gotha, Franz Xaver von Zach (1754–1832). Zach was therefore at first overjoyed at the prospect of a visit from his Parisian colleague, "my most beloved friend and grandpapa of all astronomers, Lalande," and only too pleased to grant

Lalande's request that the most prominent German astronomers be invited to Gotha to meet the celebrity guest. Despite some declined invitations (the Austrians refused to allow their astronomers to go because of ongoing hostilities with France), some fifteen astronomers from Berlin, Göttingen, Halle, and elsewhere turned up in Gotha in August 1798.

Duke Ernest II and his mother, the learned Duchess Luise Dorothea, prided themselves on being patrons of the arts and sciences, correspondents of Voltaire and other luminaries of the Republic of Letters, and were determined to pull out all the stops for the occasion. The astronomers were welcomed with a cannon salute (breaking three of the palace windows), fêted with champagne, and shown the ducal observatory's collection of top-of-the-line instruments imported from London. Zach's assistant reported with exasperation how the drunken astronomers had danced the night away and driven him half-mad with their endless questions and requests, interfering with his nightly observations. Worst of all were Lalande ("an old, vain fop who is constantly pestering me with trivia") and his niece ("an annoying chatterbox"), and after a few weeks Zach and his assistant had had quite enough of collegial partying and the grandpapa of all astronomers.

But Lalande had traveled to Gotha for more than astronomical shoptalk and champagne. The revolutionary government in France had undertaken a monumental project to create a new system of weights and measures based upon the meter, defined as 1/10,000,000 of the meridian that runs through Paris connecting the North Pole and the equator, and vaunted as nature's own universal standard, fit to replace the mare's nest of local weights and measures not just in France but throughout the world. Lalande's aim in bringing together the German astronomers in Gotha had been to persuade them to adopt and proselytize for the metric system, revolutionary France's most important (and expensive) scientific prestige project. Flattered though they might have been by the attentions of their famous French colleague and their Gotha hosts, the German astronomers were having none of it. Johann Bode (1747–1826), director of the Berlin Observatory, reported afterward that while he and his colleagues had assured Lalande that they appreciated the advantages of the base-10 metric system for purposes of calculation, they saw absolutely no prospect of introducing it into public use. They were unmoved by Lalande's claims that the meter "was taken from nature itself and therefore equally suitable for all nations" and countered his suggestion that one of the as-yet-unnamed constellations of the southern sky be called the "Aréostat," in honor of another French scientific triumph, the first human ascent in a hot air balloon by the brothers Montgolfier in 1783, with their own proposal of a constellation honoring a German invention, the printing press. Opposition was even stiffer to Lalande's advocacy of the revolutionary calendar, with its ten-day weeks and New Year's Day set to the autumnal equinox as observed from the Paris Observatory. What Lalande touted as nature's own universal system of measurement and timekeeping, his German colleagues regarded as a piece of French revolutionary propaganda—as did several contemporary newspaper accounts, which described the Gotha gathering as the meeting of a political cabal. No wonder Zach chose to remember what later historians would single out as the first international scientific congress as just an innocent gathering of friends.

Lalande's abortive mission to Gotha showed the limits of the Republic of Letters' diplomacy—and the dangers of conflating Enlightenment cosmopolitanism with ideological internationalism. Whereas mid-eighteenth-century savants largely regarded wars as a nuisance, slowing down correspondence and interfering with the book trade, late-eighteenth-century revolutionary hostilities enlisted savants in the fray. Patronage from the likes of George III or Karl Theodor depended on the whims of the monarch, but no ideological strings were attached. Support from embattled republican governments like France was often even more lavish, but only for projects judged to promote national interests and revolutionary values. As we have seen in the case of Lalande's pitch for the French metric system, national governments sought international partners to promote national prestige. By around 1800, scientific collaborations that spanned countries and oceans seemed even more improbable than they had in 1750, although communication and transportation had improved markedly. Any prospect of international governance in science seemed positively phantasmagoric. To the savage competition that had always threatened to tear the Republic of Letters apart from within were now added ideological pressures from without. The forces of distance had won out over those of proximity.

It is easy to judge the scientific Republic of Letters as a failure by the standards of later centuries. Where are the sturdy collaborations like the Laser Interferometer Gravitational-Wave Observatory (LIGO), in which astronomers from over a dozen countries banded together to detect gravitational waves for the first time in 2015; the swift diffusion of the latest laboratory results by email and online preprint archives; the scientific societies that regularly bring together colleagues from all over the world at their meetings? The gap between eighteenth- and twenty-first- (or even nineteenth-) century technologies of transportation, communication, and scientific instrumentation is certainly part of the explanation, but only part. The most sophisticated technology cannot make up for a lack of will to collaborate and a lack of stamina and loyalty to see the collaborations through. These crucial preconditions for collaboration had to be cemented by face-to-face encounters and friendships. It was personal contacts that initiated and sustained both the transits of Venus and the Mannheim meteorological observation networks for as long as they lasted. But however necessary, personal ties were never sufficient and always fragile, as Lalande's failed mission to Gotha shows.

Especially for the sciences of astronomy and meteorology, which study phenomena that transcend merely human scales of time and space, collaborations require long-term commitments and intricate fine-tuning among dispersed participants. Neither the academies nor princely courts could guarantee such continuity in the eighteenth century—and the challenge to science financed by modern democratic governments with short election cycles is, if anything, even greater. The fate of the US Superconducting Super Collider, canceled by Congress after construction had already begun in Texas, stands as an object lesson that the problem of the continuity of large-scale scientific projects has yet to be solved. Perpetuating scientific collaborations over time requires institutions that can outlast the energies and friendships that die with the initiators. No wonder so many scientists involved in such collaborations, from the seventeenth through the twentieth centuries, reached for religious analogies that likened their trans-generational projects to the longevity of monastic orders or the construction of medieval cathedrals over centuries. The eighteenth-century solution to the mastery of space was commercial imperialism (and, in the case of Jesuit observers, missionaries): the transit of Venus expeditions relied crucially on worldwide trading outposts and colonies of European powers; the Mannheim meteorological network in part foundered for lack of both. In the nineteenth and twentieth centuries, the connection between commerce, imperialism, and global science would only intensify.

Let us return one last time to the three scenes of attempted intellectual community with contemporaries as well as predecessors (figures 1, 2, and 3): the Dead Scholars Society communing with portraits; the academicians dissecting a duck together; the stay-at-home naturalist corresponding with colleagues near and far. These are three very different models of intellectual community, yet look again: they all take place in a library (even the dissection). The library was the place where the past met the present—and also the place where the present pinned its hopes on the future, on a posterity who would read their works and honor them as they deserved. Although the scientific Republic of Letters extended intellectual community on a geographic scale previously unimaginable, its primary dimension remained time, not space. The next chapter describes how the scientific congresses and collaborations of the nineteenth century attempted to combine the dimensions of time and space in a new model of international intellectual community.

Internationalism
Science as a World Project

2.1. Introduction: Thinking Globally

Circa 1800, an Atlantic crossing between New York and Liverpool with a sailing ship took between thirty and sixty days, depending on season and weather. By 1850, a steamship crossing along the same route took eleven to fourteen days, by 1875, only about eight days, and by 1900, five days. Perhaps more than any other, this statistic captures the preconditions for scientific internationalism in the nineteenth century. And not just scientific internationalism, but all kinds of internationalism, from the earliest internationalist congresses, like the World Anti-Slavery Convention convened in London in 1840, to the world expositions that drew thousands of foreign visitors to cities jostling to become global capitals of commerce, technology, culture, and tourism (figure 5). A survey of international unions in 1911 counted over 150 of them, devoted to everything from women's suffrage to railway networks to geodesy, all made possible by the acceleration of transportation and communication that united the world: "The same political dramas evoke our interest, the same catastrophes compel our sympathy, the same scientific achievements make us rejoice, the same great public figures people our imagination." Like the campaigns for an international labor union or standardized time zones or a universal bibliography, science in the latter half of the nineteenth century became a world project.

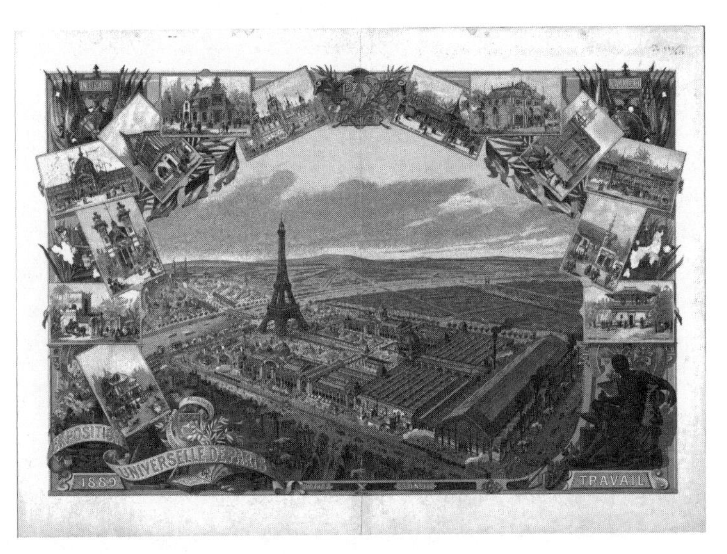

Figure 5

This is also the moment when scientists began to imagine the planet as a unified system of forces, from the great ocean currents that swirled around the earth's continents to the variations of the earth's magnetic field to the biogeography of the worldwide distribution of volcanoes and earthquake zones. In contrast to the maps of the eighteenth-century astronomers and meteorologists in chapter one, which used the world map as a backdrop for scattered observation stations, the nineteenth-century scientific maps showed the earth as an interconnected whole of invisible forces (figure 6). This brand of science is sometimes called "Humboldtian science" in honor of Prussian naturalist Alexander von Humboldt (1769–1859), who made such maps famous in his publications of his results of his many voyages to the Americas, Central Asia, and Russia. Yet Humboldt's scientific achievements showing global connections in nature were insufficient to create lasting international connections among scientists. Right up until his death in 1859, von Humboldt managed his far-flung network

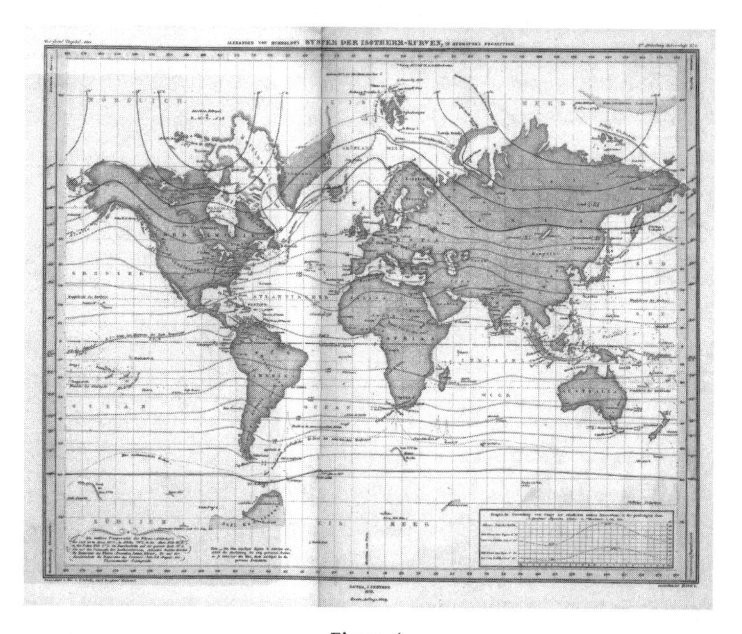

Figure 6

of scientific correspondents in the same pen-pal mode that an eighteenth-century member of the Republic of Letters had.

How, then, did the internationalism in science so much in evidence in the numerous congresses and collaborations in the latter half of the nineteenth century come about? From the historian's perspective, particularly the perspective of a historian of early modern science, nothing was less inevitable than effective global governance in science. Since the seventeenth century, academies had been founded in leading metropolises, such as the Royal Society of London (established 1660), the Paris Royal Academy of Sciences (established 1666), or the Imperial Academy of Sciences in St. Petersburg (established 1724), and by the early nineteenth century, more inclusive national scientific societies such as the Versammlung deutscher Naturforscher und Ärtzte (Society of German Scientists and Doctors, established 1822) and the British Association for the Advancement of Science (established 1831) were convening regular meetings to bring together all those interested in science, specialists and the lay public alike.

However, although these academies and societies exchanged proceedings and welcomed foreign visitors, collaborations were few and far between—as few and far between as the transits of Venus, which occurred in pairs at intervals of over a century. The worldwide expeditions to observe the transits of 1761 and 1769 were marked more by rivalry than by cooperation; those of 1874 and 1882 involved somewhat more international coordination but were still organized along strictly national lines. Researchers circa 1850 read each other, corresponded with each other, and sometimes even visited each other. Most read at least one other language besides their own. But they did not collaborate with each other across national boundaries, much less submit themselves to binding international agreements that obligated generations to come. Science circa 1850 was global in its scope and cosmopolitan in its networks, but its governance was at best national, and even that was precarious.

Fast-forward to around 1900, and the sciences (and some of the humanities) had become almost as global as the phenomena they studied. Delegates from over a dozen countries met in Paris in 1872 to thrash out the technical specifications of new standards for the meter and kilogram; chemists gathered in Geneva in 1892 to agree upon a standard nomenclature scheme for newly synthesized compounds; botanists convened in Vienna in 1905 to hammer out rules for botanical nomenclature and claiming priority for the discovery of new species.

All of these meetings were tense with controversy. Much was at stake: professional reputations, commercial interests in the manufacture of compounds and instruments, institutional interests in the management of specimen collections, research agendas for years or even generations to come, and national cultural prestige on the world stage. Delegates collided time and again over these issues, as the minutes of these meetings reveal. Yet in the end, resolutions were passed and, more significantly, honored for decades, despite the disruptions of war, revolution, decolonialization, and the complete remaking of the geopolitical order in the course of the twentieth century.

Even more impressive and improbable were the long-term, labor-intensive, and expensive international scientific collaborations initiated during the same period, such as the Carte du Ciel begun in 1887 or the Internationale Erdmessung (International Geodetic Association) begun in 1886. As their names indicate, the initiative for these projects had emanated from one nation (France and Prussia, respectively), but their global dimensions required a worldwide network of observers. These cooperations involved intricate negotiations over every detail in order to standardize the results produced by participants scattered all over the globe, often working under challenging field conditions, as well as constant fundraising from patrons, both private and public, to foot the ever-larger bills for the research. Most significantly, international scientific cooperations involved the very real sacrifice of individual careers and local research priorities to the interests of an international collective—and not even a collective of colleagues in the here and now, but one imagined in the distant future.

It is therefore a double wonder that these collaborations ever came about, much less endured. First, science was just as savagely competitive in the latter half of the nineteenth century as it is now. Although it was no longer necessary for scientific academies expressly to forbid name-calling and fisticuffs at their meetings, as the Paris Royal Academy of Sciences had in its regulations of 1699, fierce competition and scathing polemics in print were still the rule rather than the exception in nineteenth-century science. One reason why international congresses of botanists, chemists, zoologists, meteorologists, and other disciplines were convened with increasing regularity in the latter half of the nineteenth century was to lower the temperature of professional quarrels by in-person sociability, often lubricated by barrels of booze and after-hours drinking songs composed specially for the occasion.

Second, the period between 1840, when the first international congress was held in London by advocates of the abolition of slavery, and the outbreak of World War I in 1914 was one of intense national military, imperial, economic, and cultural rivalries as well as of international projects, and science was no exception. Like so many other starry-eyed world projects of the period, the scientific collaborations, always fragile and underfunded, were menaced by national hostilities, which could range from a declaration of war of the sort that capsized Franco-Prussian relations after 1870, to the prickly sensibilities of British astronomers defending the superiority of Newtonian reflecting telescopes over French refractors.

And yet, wonder of wonders, an astonishing number of these international collaborations did survive—and moreover laid the foundations of current international scientific governance. Scientific phenomena have always been global, but global science emerged only in the latter half of the nineteenth century: How did this come about, in the teeth of the odds?

This chapter concentrates on the early stages of two successful projects, both still in use today, and both once again taken from astronomy and meteorology, in order to sharpen the contrast with the examples from chapter one: the international astrophotographic star chart, known as the Carte du Ciel, and the International Cloud Atlas, first issued in 1896 and most recently updated in 2017 (as an online edition). In both cases, these ambitious projects paved the way for the creation of international governing bodies for their disciplines, the International Astronomical Union (established 1919) and the World Meteorological Organization (established 1950).

But these two cases are also a study in revealing contrasts. No one has ever doubted the global dimensions of star maps, but many late-nineteenth-century meteorologists entertained serious doubts as to whether the same classification held for clouds all over the world. As its name indicates, the Carte du Ciel was a French prestige project, backed with generous government funding and all the glitter and glamour of Paris, whereas the International Cloud Atlas was bankrolled by a private patron and meetings were held at the editor's home in Uppsala, Sweden. The Carte du Ciel aimed to provide a tool for the astronomers of the future, while the International Cloud Atlas was pressed into immediate service by observers all over the world. Yet both collaborations faced the same challenge: how to make resolutions adopted by a handful of practitioners binding on the entire discipline for generations to come, regardless of drastically changed circumstances—a challenge that no diplomatic treaty among sovereign nations has yet met.

2.2. A Universal Parliament

When champions of world peace, world government, even world statistics around 1900 looked for inspiration and proof that their projects were practicable, their sterling example of success was neither an international diplomatic treaty nor an international scientific collaboration nor even an international commercial partnership. Such agreements were one-shot deals, agreed upon for the mutual convenience of the signatories and only as enduring as the convergence of those interests. No, the paragon of successful international governance, the one utopian scheme that really worked, was the Universal Postal Union, established in Bern in 1874. Even in the thick of World War I, idealists pinned their hopes on a world government emerging from the carnage of the trenches on the example of the Universal Postal Union. Here is the British Fabian Leonard Woolf praising the Postal Union as the harbinger of world peace and the end of bellicose nationalism at a moment when most of the world was at war:

> The Universal Postal Union after a life of over forty years remains the most complete and important example of international administration [. . .], having by its birth effected a revolution in the constitution of the society of nations.

Others celebrated the Postal Union as "a universal parliament" and as the most enduring and inclusive of all international agreements, a milestone in the history of international law.

Prior to the creation of the Postal Union, international postal exchanges had been regulated—if they had been regulated at all—by a cat's cradle of bilateral treaties. Depending on which route a letter was sent, say, from Munich to New York, it could cost anywhere between 4 kroners (sent via Bremen) to 10 (sent via Cologne and England). A letter sent from New York to Australia could travel any one of six different routes, costing anywhere between 5 cents and $1.02 in postage. The rules governing which rates countries along each route could charge in additional postage were even more labyrinthine, not to mention the conversions among different currencies and different systems of weights and measures. When in 1874 delegates from 21 countries signed the Treaty of Bern (representing 192 countries as of 2011, when the newest member joined) creating a uniform postage rate of 25 centimes for letters under 15 grams mailed anywhere within the territories of the signatory nations, it seemed to contemporaries tantamount to the ex nihilo creation of order out of chaos. When, seventy-five years later in 1949, the Universal Postal Union in Bern was still quietly and efficiently regulating the world's postal exchanges after a horrendous world war that had shattered almost all other international treaties, it seemed a beacon of functioning international cooperation amidst a storm of national hostilities.

Quite aside from its longevity and efficacy, three aspects of the Universal Postal Union provided a template for subsequent international scientific collaborations.

1. *The initiative came from high-level postal officials, not from their governments.* At the personal request of US Postmaster Montgomery Blair (1813–1883), postmasters from fifteen countries met in Paris in 1863 to try to sort out the international postal mess. Although their local host, French postal director Édouard Vandal (1813–1889), sternly reminded the delegates that they had no authority to commit their governments to any binding agreements—that was the purview of diplomats—the postmasters in attendance managed to hammer out thirty-one resolutions on matters of principle. These and the proposals circulated by the Prussian Ober-Postrath Heinrich von Stephan (1831–1897) in 1868 formed the basis of the 1874 Bern congress, which eventually did result in a treaty.
2. *The delegates were all postmasters, not professional diplomats.* They all faced similar problems in their home countries, and the experience of several weeks of collegial shoptalk in the intoxicating atmosphere of Paris and even the not-so-intoxicating atmosphere of Bern

created lasting personal bonds that smoothed even the most difficult discussions—for example, as to whether colonies should be allowed to vote independently—and paved the way for behind-the-scenes compromises.

3. *They created an international bureau to do the real work.* Although the international congress of plenipotentiaries sent by signatory nations met at regular intervals, the real work of applying and tweaking the rules, preparing resolutions for upcoming congresses, and making occasional exceptions was carried out by a permanent international bureau located in Bern—not the capital of a great power like France or Britain, but of a small, neutral, and, crucially, polyglot country. Although the official language of the Postal Union was French (a sop thrown to the balky French delegates), bulletins were published in French, German, and English. It was the permanent, multi-lingual staff of the Bern bureau, not the grand but unwieldy congresses in Paris (1878, 1880) or Washington (1897) or Rome (1906), that kept the Postal Union functioning in "placid obscurity" like well-oiled Swiss clockwork.

The moving spirits behind the Carte du Ciel and the International Cloud Atlas learned these lessons well. *First*, begin with informal gatherings of specialists and formulate resolutions of principle before soliciting government approval and support for putting principles into practice; *second*, cultivate collegial sociability in amiable surroundings in order to establish trust, willingness to compromise, and, above all, long-term loyalty to the project across national boundaries; and *third*, create a permanent committee or bureau, preferably led by a linguistically gifted representative of a small country, to oversee the day-to-day workings of the project in all their knotty detail. The template for internationalism provided by the Universal Postal Union tempered national sovereignty by specialist collegiality and encouraged backstage compromises on national priorities. It aimed at internationalism without nations—except when it came time to pay the bill.

2.3. The Carte du Ciel: The Diplomatic Model

The April 1887 gathering of the world's astronomical elite at the Paris Observatory to plan the astrophotographic map of the sky was an extravaganza. Decked out in borrowed finery of the sort usually reserved for state occasions, the Observatory glistened with silver candelabra, Sèvres porcelain, and gilded Louis XIV armchairs. Nine-course banquets and evening concerts leavened the long days of deliberations on whether reflecting or refracting telescopes were best suited to astrophotography, how to divide up the labor of photographing the whole sky among the eighteen participating observatories, and which emulsion to use on the photographic plates. Admiral Ernest Mouchez (1821–1892) and subsequent directors of the Observatory staged the Carte du Ciel with all the pomp and circumstance of a diplomatic congress, tying the project's success to national glory. Whenever the French government balked at the mounting expenses, the Observatory director countered that the success of the project was "a point of honor for France" (figure 7).

The fifty-eight astronomers from sixteen countries plus three colonies who met in Paris planned what one contemporary called "the greatest venture yet undertaken in astronomy,"

Figure 7

namely a complete photographic map of the sky, including all stars to the fourteenth magnitude, made possible by the new astrophotographic techniques pioneered by Edward Pickering (1846–1919) at Harvard and the brothers Paul (1848–1905) and Prosper Henry (1849–1903) at the Paris Observatory.* Only the combined and prolonged efforts of almost a score of observatories in both the northern and southern hemispheres

* A star catalogue down to the eleventh magnitude was also planned as part of the Carte du Ciel project. On the history of nineteenth-century astrophotography in general, see Daniel Norman, "The Development of Astronomical Photography," *Osiris* 5 (1938): 560–594; Dorrit Hoffleit, *Some Firsts in Astronomical Photography* (Cambridge, MA: Harvard College Observatory, 1950); John Lankford, "The Impact of Photography on Astronomy," in *Astrophysics and Twentieth-Century Astronomy to 1950*, ed., Owen Gingerich (Cambridge: Cambridge University Press, 1984), 16–39.

could produce what promoters hailed as an "imperishable monument," a photographic record of "the authentic state of the universe visible from the earth at the close of the nineteenth century." Eighteen observatories around the world, from Helsinki at +60.9 degrees latitude to Melbourne at −37.5, would join forces to create the most complete star map in the history of astronomy, a legacy for astronomers in the year 3000 and beyond (figure 8). The delegates to the Carte du Ciel meetings hoped that future astronomers would benefit from their snapshot of the sky as seen from the earth circa 1900 to transcend the short span of a human life, or even of human generations, detecting the appearance of new stars, nebulae, and comets, the telltale motion of as yet undiscovered planets, the extended periods of variable stars, and the incremental proper motions of the so-called fixed stars.

As the deliberations of the 1887 International Congress and of subsequent meetings (1889, 1891, 1896, 1900, 1909) of the Permanent Committee make clear, the intricate coordination of telescopes, photographic plates, micrometric measurements, and myriad other details in order to insure that the parts of the map would be commensurable required that participants relinquish control not only over instruments and methods, but also over the choice of research area for decades to come. The levels of sacrifice demanded by the scientific collectivity were enormous: the cost in time and money of new instrumentation and training, the substitution of efficiency for painstaking precision, the monopolization of resources and personnel for long periods of routinized labor, the steadfast resistance to the temptation to neglect old collaborative commitments in pursuit of an exciting new discovery. Some observatory directors, including

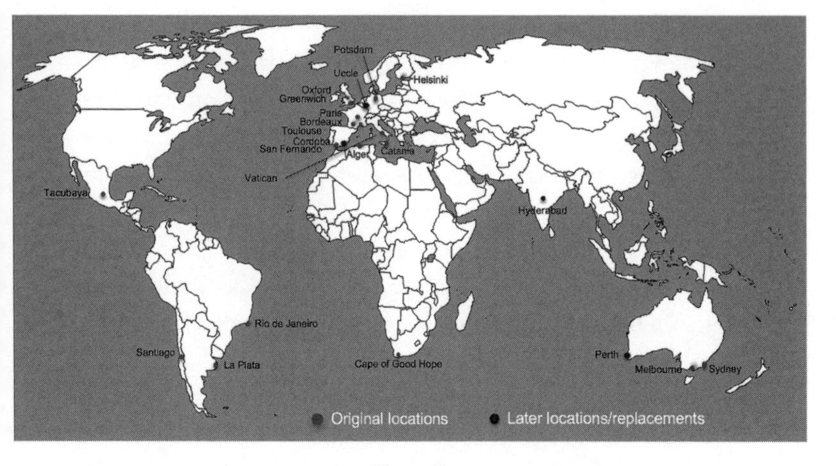

Figure 8

T. N. Thiele of Copenhagen and Pickering at Harvard, judged the costs of collaboration to be too great and declined to participate. The sacrifices demanded of those who did participate were particularly grave for the smaller observatories: for example, the Australian observatories of Sydney, Perth, Melbourne, and Adelaide took eighty years to complete their three assigned zones of the sky (18 percent of the entire sky), at the price of limiting other investigations, particularly in the enormously fruitful fields of astrophysics and spectroscopy that came to dominate twentieth-century astronomy.

The Carte du Ciel collaboration, which was ended (not completed) in 1970, is an example of a scientific collaboration that began just as the Postal Union had, as the personal initiative of one or two directors of national institutions—Admiral Ernest Mouchez and David Gill, directors of the Paris and South African observatories, respectively—but from the outset strove to emulate at least the trappings of a full-dress diplomatic congress. Invitations were sent out by diplomatic pouch; the Paris Observatory was given a face-lift for the occasion with bling requisitioned from the Mobilier National; Mouchez and subsequent directors of the Paris Observatory made French leadership of the project a matter of national pride—and also a goad to their British colleagues to raise comparable sums to finance the mounting costs of publishing the photographic plates. When Benjamin Baillaud (1848–1934), Mouchez's successor as director of the Paris Observatory, presented the French minister of education with a whopping bill of 1 million francs for the copper plates needed to engrave the photographs of the stars, he simultaneously vaunted the value of the plates as an "inalterable . . . [and] rigorous inventory of a part of the sky at the beginning of the twentieth century" left to "the generations to come," and shrewdly remarked that if the French government shelled out—yet again—for half the costs, it might shame the British into paying the other half.

It was also a matter of French prestige that the Permanent Committee formed to adjudicate delicate matters such as the photometric scale for assigning luminosity magnitudes to stars be centralized in Paris—rather than in some more modest, neutral capital like Bern, although Otto von Struve (1819–1905), the suave, polyglot director of the Russian observatory at Pulkowa, played a crucial diplomatic role behind the scenes and was made a member of the Légion d'Honneur for his troubles. But however much especially British national sensibilities might have smarted under French leadership, the anglophone delegates were well aware that this was the condition for hefty French government subsidies for the never-ending project. For the French government, lavish hospitality to scientific congresses and generous aid to scientific projects under French leadership was part of a concerted strategy after 1871 to compensate for military losses to the Germans and commercial and imperial losses to the British by making Paris the cultural capital of the world. Meetings of the Carte du Ciel congresses after 1887 were deliberately timed to coincide with world expositions in Paris, like that of 1900, so that foreign delegates (and their families) might admire the city of light at its brightest.* French savants and diplomats, often working in concert with one another, mastered the art of bringing together the world's scientific elite in Paris, wining and dining them in grand style and recruiting them into scientific collaborations more binding than treaties and often more lasting than nations and empires.

2.4. The International Cloud Atlas: The Voluntarist Model

The project to create an International Cloud Atlas launched at the Munich meeting of the World Meteorological Conference in 1891 enjoyed none of the quasi-diplomatic pomp and circumstance of the Carte du Ciel. Far from being welcomed by local dignitaries and fêted at sumptuous dinners, the delegates to the Munich congress were not even sure that the Bavarian regime would allow them to meet, much less roll out the red carpet. As British delegate Robert Scott (1833–1916) reported to the Royal Meteorological Society in London, the gathering had been demoted from "congress" to "conference" because the

* After 1889, the French government engineered international congresses to coincide with the universal expositions held in Paris. During the 1900 universal exposition, 122 congresses were held in Paris. John Culbert Faries, *The Rise of Internationalism* (New York: W. D. Gray, 1915), 41.

 attendees were "not representatives of governments" and participated in a meeting that deliberately described itself as "more or less, [of a] private character." Nor did the resolution to create a committee to select typical images of cloud types in order "to introduce uniformity in the classification and nomenclature in cloud observations" command universal assent. Even among the expert meteorologists, there was disagreement as to whether, for example, the clouds of Scotland resembled those of Italy, much less those of Chile or Australia. Some wondered whether stable cloud forms that could be classified in genera and species existed at all. As even the more recent (1987) *International World Cloud Atlas* concedes, "[C]louds are continuously in a process of evolution and appear, therefore, in an infinite variety of forms." Whereas none of the astronomers doubted for a moment that a snapshot of the heavens as seen from all over the earth circa 1900 would be invaluable to future astronomers of the year 3000 and beyond, a considerable portion of meteorologists doubted the utility of a cloud atlas that aimed to guide observers all over the globe.

The one advantage the initiators of the cloud atlas project had over their counterparts in the Carte du Ciel was urgency: by 1891 it was clear that the simple categories introduced almost a century earlier by Jean-Baptiste Lamarck (1744–1829) and Luke Howard (1772–1864) were inadequate to the needs of the mariners and farmers who were the main clients of national meteorological services. Cloud classification had begun earlier in the century, with the publication of Howard's *On the Modification of Clouds* (1803). But by the 1870s, classification systems based on Howard's original tripartite scheme of cirrus, cumulus, and stratus had splintered and ramified in the prolific fashion of the clouds themselves. Worse still, names had come unstuck from the things they were supposed to designate: a Swedish, Portuguese, and British observer might all mean different things by the designation "cumulo-stratus"; observers beyond Europe diverged even more widely from one another. In principle, this pluralism of cloud classifications need not have been diagnosed as a crisis. If clouds were variable, local phenomena, why try to impose the same classification scheme on observers in Rio de Janeiro and Uppsala? If the Deutsche Seewarte and the Portuguese navy differed regarding the number and definition of cloud rubrics in which sailors were instructed, might this not correspond to the well-known differences between the weather on northern and southern seas? In practice, however, the meteorologists reacted with alarm. In their eyes, science was international because nature was universal. A global classification of clouds was hence a precondition for and a product of international scientific collaboration.

By the 1873 meeting of the International Meteorological Congress in Vienna, the situation was judged intolerable: various observatories were invited to "publish exact representations of the form of clouds considered typical at each location." When Hugo Hildebrandsson (1838–1925), director of the Uppsala Observatory, and the British meteorologist Ralph Abercromby (1842–1897) joined forces in the 1880s to correlate the major systems of cloud classification then in use, they discovered that at least three designations converged: cirrus, cumulus, and cirro-cumulus. It was these reassuring findings that emboldened the 1891 Munich conference to endorse the Hildebrandsson-Abercromby cloud classification system and to produce an atlas on that basis, to be tried out in practice by observers all over the world in 1896, designated as the International Cloud Year.

The members of the Atlas committee met in Uppsala in August 1894 to choose among some three hundred images (figure 9). These included not only photographs, but also paintings and pastels; the Danish artist and cloud-observer Philip Weilbach had also been invited to participate in the committee's deliberations. Clouds were divided into ten numbered types, further subdivided by altitude and whether they portended good weather or bad. Each of the main types as well as some secondary types were assigned a representative image. What qualified as a representative image was a matter of collective judgment; no single person and no single medium held a monopoly. While the majority of the twenty-eight plates were photographs, paintings and pastels were also included. Hildebrandsson and

Figure 9

Abercromby were both on record as proponents of cloud photography, because of the detail captured in that medium, but both also acknowledged limitations. Hildebrandsson insisted that color was essential. He had for this reason employed artists who worked from nature or from "good photographs" to produce paintings and chromolithographs for the 1890 cloud atlas he had published with the Hamburg meteorologists Georg von Neumayer and Wladimir Köppen, who had also employed an artist. Here as well the choice of images had been collective; Hildebrandsson had traveled to Hamburg in 1888 expressly for this purpose and then had some of the paintings redone on Weilbach's advice.

Hildebrandsson was the quintessential scientific diplomat of the kind who so often figured in the origins of successful international scientific collaborations: of sufficient professional stature to command the respect of colleagues from the bigger countries (France, Germany, Britain) but hailing from a more peripheral country (Sweden); able to speak and correspond effortlessly in the three major languages of the International Meteorological Conference; warmly hospitable as the committee members and their families congregated in Uppsala; and indefatigable in pursuit of the committee's declared aim of producing the atlas in time and according to the specifications set down by the committee—even though it almost certainly meant that his own 1887 atlas, published with Abercromby, would be thereby instantly outdated. Thanks in no small part to Hildebrandsson's efforts, the bonds cemented during the August 1894 meeting in Uppsala created a reservoir of trust and commitment that was as personal as it was professional. Hildebrandsson's papers in Uppsala are strewn with thank-you notes, fond reminiscences of the August spent in Uppsala, and family news sent by committee members for years afterward.

By coordinating their own judgments of typicality, the Atlas committee hoped to make it possible to coordinate those of observers worldwide. The aim of presenting secondary as well as primary cloud forms was to "direct thereby the attention of observers to the characteristic differences between these [primary] types and the forms derived from them." At the same time they selected images together in Uppsala, the Atlas committee also fixed the definitions and descriptions, as well as the observing instructions: further grids for the calibration of perception, again modeled in microcosm by the collective activities of the committee itself. The consensus of the committee was the only authority backing up a classification system that was intended to guide cloud observation forever after—not unlike the authority projected, again from Uppsala, by Carolus Linnaeus's (1707–1778) botanical classification system over 150 years earlier, the *Systema naturae* (1735).

Just how seriously the committee took the resolutions reached collectively in Uppsala in August 1894 is made clear in the correspondence of the three members charged to publish the atlas: Hildebrandsson in Uppsala, Léon-Philippe Teisserenc de Bort (1855–1913) in Paris, and Albert Riggenbach (1854–1921) in Basel. Endless difficulties plagued the production of the atlas: What printing process would do justice to the all-important images and yet be cheap enough for observers all over the world to buy? Should the printing be done in Zurich (top quality but very slow) or Paris (faster but not so attentive to the all-important illustrations)? Who would pay for the mounting extra costs? Meanwhile, the clock was ticking. As Hildebrandsson frantically admonished Teisserenc de Bort (in French) and Riggenbach (in German), observers all over the world were waiting to receive and try out the new atlas in the International Cloud Year of 1896: "on all sides I hear cries and demands to send the atlas!"

Yet even when speedy publication or the quality of an image was at stake, the three editors felt the decisions of the Uppsala committee to be binding. Riggenbach much preferred the photograph Hildebrandsson had sent him of a nimbus cloud to the pastel rendition chosen by the committee, but both decided that they had no authority to alter the decisions of the committee. Princess-and-pea national sensibilities were treated with exquisite tact. The original meeting of the Atlas committee was postponed by a year so as not to offend the American members attending the 1893 meeting of the International Meteorological Congress in Chicago; advance copies of the chosen images were sent to tetchy German colleagues offended because their atlas hadn't been chosen; endless pains were taken to make sure that the text and especially the title translated smoothly into all three languages. Even though Tesseirenc de Bort ended up saving the atlas by contributing privately to the costs of publication, his pet French title, *Atlas-Normal des nuages*, was sacrificed because the title *Atlas international des nuages* better reflected the international resolution that had brought it into existence. A voluntary association lacking both official recognition and support, the committee that assembled the first International Cloud Atlas clung to their one claim of legitimacy, namely that they represented an international consensus of their colleagues.

2.5. Conclusion: Cooperate or Defect?

However different in their objects and models of collaboration, both the Carte du Ciel and the International Cloud Atlas were creations of their epoch. Neither star maps nor cloud classifications were new in the late nineteenth century, but the then-novel technology of photography fired the imagination of both the astronomers and meteorologists. Like many other world projects of the latter half of the nineteenth century, most of the lasting achievements of scientific internationalism involved standardization, whether of weights and measures or botanical nomenclature or photometric indices of stellar luminosities. Some of the scientific projects went further and attempted to standardize instruments, methods of measurement, and even times of observation. The scientific movement to standardize and synchronize mimicked and, in the case of weights and measures, overlapped with commercial impulses to smooth the way for world trade in a period when exports and imports to Europe alone increased over forty-fold. Like commercial internationalism, science benefited enormously from speedier transportation and communication. Without the steamship and telegraph networks that accelerated international contacts and without the imperial possessions of European powers that provided observational footholds around the world, neither collaboration would have been conceivable. And like the huge boom in international trade during this period, 75 percent of which involved industrialized nations (not their colonies), scientific internationalism was centered on Europe and North America.

Yet despite the special circumstances that facilitated the Carte du Ciel and the International Cloud Atlas, both survived their founding moment and became the basis for their discipline's international governing body, despite decolonization and world wars that disrupted communication and transportation. The International Cloud Atlas has been revised and reissued in multiple editions, most recently in 2017, and continues to guide thousands of observers worldwide. The Carte du Ciel endured a long fallow period during which the big observatories like Paris lost interest but the small ones in the provinces and colonies soldiered on, ultimately becoming a commission of the International Astronomical Union. After it was summarily ended in 1970, work on the never-completed star chart stopped, and in observatories all over the world the around 2 million glass photographic plates gathered dust in file cabinets. Then around 1990, this Sleeping Beauty archive awoke. By comparing the positions of the Carte du Ciel catalogue with those of the new Tycho catalogue made with data from the European Space Agency Hipparcos satellite, it was in fact possible to calculate the proper motions of almost a million stars, just as Mouchez and the other initiators of the Carte du Ciel had hoped.

But these successes could not have been foreseen in the darker days of the very dark century that intervened between the optimistic internationalism that had propelled both projects at the end of the nineteenth century. And even before two World Wars, the Great Depression, multiple revolutions, and the making and remaking of the geopolitical order in the twentieth century, there had been discord aplenty behind the scenes of both projects. The participants of the Carte du Ciel squabbled over everything from which telescope to use (the French refractor or the British reflector?), which grid to use in measuring the photographic plates (the British loathed the German suggestion), which observatories would be saddled with the drudge work of measuring the plates (here the French provincial and colonial observatories won the booby prize), and, most contentious of all, what would be the language of publication (a compromise proposal that authors publish in their own languages won out, over loud French protests in favor of the priority of their own language, as host nation). The committee charged with putting together the first International Cloud Atlas also occasionally fractured along national lines (translations of the proceedings of meetings of the International Meteorological Conferences were done by volunteers among the attendees, who complained about the extra workload) and even more along lines of professional judgment. Hildebrandsson privately abhorred the pastels of clouds inflicted on the Uppsala committee by the Danish artist Weilbach but felt obliged to include them because the committee had so ruled. It is astonishing that these projects ever got off the ground, much less that they survived loss of funding, the shredding of the international order, and waning professional interest as attention shifted to the latest hot research topic.

There were certainly defections along the way, and work on both projects often slowed to a standstill during wars, revolutions, depressions, and plain old staffing shortages. But somehow the projects limped on. Why? In the language of game theory, why did the people and institutions charged with completing the star map and updating the cloud atlas decide to cooperate rather than defect? The question is all the more pointed in light of the fact that some *did* defect, without suffering any substantial sanctions. The reason why the Australian observatories ended up mapping so much of the southern sky was that almost all of the other observatories in the southern hemisphere, most in former European colonies, dropped out over the course of the twentieth century. To repeat the game theorist's question: Why cooperate rather than defect?

This question seems most pressing for the Carte du Ciel, a long-haul project that lasted over eighty years, apparently showing no returns for much of that period. But the fate of these projects in the short haul was even more precarious: there were not yet sunk costs in time, money, and reputation to serve as a check to the self-interested urge to jump ship. Moreover, aside from the example of the Postal Union and diplomatic congresses, both ultimately resulting in treaties among sovereign nations, there was no road map for international scientific collaborations. There were not even rules to regulate how to conduct meetings. One of the British delegates to the 1887 Carte du Ciel meeting, Oxford astronomer Herbert Hall Turner, reminisced wryly in 1912 about the pandemonious early meetings: "The discussions were, to say the least of it, animated. There are no universal rules for conducting public business, and astronomers of one country were not familiar with rules in use elsewhere." A common culture had to be established before delegates could even make themselves heard above the din of conflicting opinions shouted out in multiple languages.

That common culture was borrowed from motley sources, including the protocols of national academies and specialist scientific societies as well as the rituals of diplomatic congresses. But this patchwork of borrowings alone would not have sufficed to create the trust, goodwill, and loyalty necessary to win commitment of substantial resources, much less to sustain the cooperation through hours and hours of rancorous deliberations about the details of photographic emulsions or the best image of a stratus cloud. The glue came from what might at first seem to be the flotsam and jetsam left by those early meetings in the archives: the dinner menus, the programs for the soirées musicales, the elaborate toasts exchanged between hosts and guests, the thank-you notes for private dinner invitations and chatty exchanges of collegial gossip. Scandalized though he was by the French practice of having the chairman move motions in meetings, Turner warmly praised French "cordiality and hospitality that has never failed to impress their colleagues from the most distant parts of the world." British meteorologist Robert Scott, who made a point of attending all the international meteorological meetings, even though he complained about having to translate the proceedings of the morning's meeting into English during the afternoon, singled out the bonhomie that prevailed as his main motivation for going. "The whole proceedings were marked by the most thorough good fellowship, and, in fact, it is in the fostering of this feeling, much more than in the discussion of abstruse scientific questions, that the real value of these international gatherings is to be found."

The most successful of these gatherings were also the most exclusive. The participants in the meetings that laid down the law on how to name new chemical compounds or zoological species or unusual cloud formations all over the world and for generations to come were the elites of their fields and almost to a man, men.* In the case of the Carte du Ciel and the Cloud Atlas

* At least in the case of the Carte du Ciel, women played an important role behind the scenes measuring the photographic plates. Observatories at Paris, Harvard, Oxford, and elsewhere made use of the first generation of women to benefit from university education in astronomy but who could be paid lower wages than their male counterparts for much of the

projects, they were directors of observatories and national or regional meteorological services, prominent figures with a reputation and resources who had already met in the pages of specialized journals before they met in person in Paris or Munich. In the language of male honor that saturated so much of the after-hours conviviality at such gatherings, they regarded each other as peers, capable of giving satisfaction in a scientific duel but also of upholding a gentleman's agreement guaranteed by no legal document. Like the postmasters who met in Paris and Bern, they reveled in the opportunity to discuss technicalities that only a fellow-aficionado could love for days on end. Whereas the Republic of Letters had been wary of the corrupting influence of proximity on scholarly and scientific judgment, preferring the verdict of foreigners and posterity to that of compatriots and contemporaries, late-nineteenth-century scientific internationalism thrived on these occasional but intense bursts of collegiality, often combined with tourism at a conveniently timed world exposition nearby.

Those all-too-familiar group photos of both cooperations are worth a second look (figures 7 and 9). Our eyes are blind to the symbolism of such conference photos—we've seen too

labor-intensive work in astrophotography and later spectroscopy: Pamela Mack, "Straying from Their Orbits: Women in Astronomy in America," in *Women in Science: Righting the Record*, eds., G. Kass-Simon, Patricia Farnes, and Deborah Nash (Bloomington: Indiana University Press, 1996), 72–116. The only woman present at the 1887 Carte du Ciel banquet (besides the official hostess, Mme Mouchez) was the American-born Dorothea Klumpke (1861–1942), the first woman to receive a doctorate in astronomy from the Sorbonne and head of the all-female bureau of calculation at the Paris Observatory: John Henry Reynolds, "Dorothea Klumpke Roberts," *Monthly Notices of the Royal Astronomical Society* 104 (1944): 92–93.

many of them (even posed for a few ourselves), all of them boringly similar in their stiff rows of bearded gentlemen in bowlers and top hats. But these two come from a moment when the idea of the international scientific meeting was a new and untried idea; indeed, the very idea of an international scientific community was crystallizing. It was a community that required face-to-face encounters, preferably over a convivial nine-course dinner, to coalesce. The language the people in these photos used to describe this community was a strange mixture of stern appeals to duty, fervid invocations of religious sentiment displaced to science, male honor, and slightly tipsy camaraderie. Whatever the register, devout or jolly, the theme of sacrifice recurs over and over again—the sacrifice of national pride in language and research tradition, the sacrifice of one's favorite telescope or cloud photo, the sacrifice of short-term individual scientific interests to those of the long-term collective. What these photos commemorate is not just who was there in Paris in 1887 or in Uppsala in 1894; they commemorate the will to bring the scientific community into existence, meeting by meeting, dinner by dinner.

The Scientific Community
Governance without Governments

3.1. Introduction: The Cosmic Community

In September 1875, the American philosopher and physicist Charles Sanders Peirce (1839–1914) attended an international conference in Paris on how to determine the shape of the earth. The Gradmessung, as it was called, was one of those nineteenth-century international scientific collaborations described in chapter two. Originally launched by the Prussians in 1864 as a Central European measurement network, by 1875 it had become a full-fledged international collaboration—literally a world project that sought to ascertain the true shape of the earth by measuring the value of the gravitational constant at scattered points across the globe using carefully calibrated Repsold pendulums. Peirce had been sent to the Paris conference as a delegate from the US Coast and Geodetic Survey, and he entirely subscribed to the combined ethos of precision and coordination that characterized the Gradmessung collaboration in geodesy as much as it had the Carte du Ciel in astronomy. At the Paris conference he suggested that all observers bring their pendulums to Berlin to have them calibrated there, as Switzerland and Austria already had, and he was scathing about those countries that refused to do so. He accused the holdouts of being "willing to sacrifice the solution of a great problem to forms of action based on national exclusiveness," in defiance of the "international solidarity" required for the scientific project to succeed.

Peirce's hands-on experience swinging pendulums in Berlin, Kew, Paris, Geneva, and Hoboken as part of an international scientific collaboration deeply imprinted his later philosophy of science and logic. Since all inductive inferences were a matter of probability, he argued, greater certainty could only be achieved by confirming them over and over again by a community of inquirers that spanned the globe and stretched across generations. Indeed, Peirce's ideal community potentially extended beyond the planet Earth and this geological epoch: "This community, again, must not be limited, but must extend to all races of beings with whom we can come into immediate or mediate intellectual relation. It must reach, however vaguely, beyond this geological epoch, beyond all bounds."

Around 1900, Peirce was not alone in his rhapsodic vision of the scientific community as not just global but interplanetary. In a series of lectures on the latest developments in physics delivered at Columbia University in 1909, Max Planck (1858–1947) had urged his colleagues to give up the "intuitions and immediacy" of human experience in order to unify the community of "physicists in all places, in all epochs, in all cultures . . . [and] also the inhabitants of other planets." Peirce came from what in those days was the scientific periphery in Washington, DC, and Planck was enthroned at its very center in Berlin, but for both the heady experience of participating in international conferences and collaborations, from the Gradmessung project to the Solvay Conferences (established in 1912/13 to bring together the world's most eminent physicists and chemists), kindled the hope of a cosmic scientific community.

This vision of an all-inclusive scientific community—indeed, the very term "scientific community"—was something of a rarity at the beginning of the twentieth century. A Google Ngram gives a rough idea of the frequency with which the term was used. "Scientific community" occurs relatively seldom until about 1950, plateauing around 1980. Delving into the texts that document the earlier uses of "scientific community" prior to 1950, one finds it mostly used in conjunction with references to the nation-state, either to the scientific community of a particular nation or colony (e.g., the scientific community of Australia) or in opposition to the government of a particular nation (e.g., the British government's neglect of the views of the scientific community). As often as not, the term is used in the plural, "scientific communities," either to refer to specialized disciplinary groups (e.g., to chemists or botanists) or to the scientists of various nations and provinces. Only after 1945 does the term take on a consistently unified and international character. What happened between the internationalist, even interplanetary euphoria expressed by Peirce and Planck circa 1900 and the belated crystallization of the idea (and arguably the reality) of *the* scientific community, embracing all disciplines and nations, after about 1945?

This timing is at first glance puzzling. Had not two devastating world wars in which scientists had for the most part been as fervently nationalistic as their compatriots destroyed internationalism in science, making a mockery of all talk of scientific community? The terrible decades between 1914 and 1945 could have well interrupted the scientific internationalism of the late nineteenth century not just temporarily but ended it altogether. Why then did the ideal of the scientific community, in the singular rather than the plural, emerge in the aftermath of the carnage? And was it ever anything more than an ideal? How did scientific internationalism survive two world wars, and was the idea of the scientific community that emerged after 1945 really the same as the internationalism that thrived at those conferences in Paris, Geneva, Brussels, Chicago, and other metropolises alongside world expositions and diplomatic galas prior to the outbreak of World War I?

This chapter argues that scientific internationalism of the late nineteenth century did *not* survive the First World War, much less the Second. Scientific internationalism circa 1900 was implicitly based on a model of friendly national competition, much as countries nowadays keep proud score of the number of Nobel laureates they produce. Because scientific nationalism went hand in hand with scientific internationalism, the surest way to extract money to support some international scientific project from a reluctant government was an appeal to national pride, a strategy that the French and British astronomers had already perfected by the late eighteenth century. As we have seen in chapters one and two, personal and national rivalries were always simmering beneath the surface of the international congresses of the late nineteenth century, but skilled polyglot diplomacy and boozy conviviality usually succeeded in making peace behind the scenes. In the bitter aftermath of World War I, however, even the most ardent prewar internationalists, most notably the Belgians, insisted upon excluding scientists from the defeated Central Powers from international scientific organizations. The next section uses the case of the ill-fated International Association of Academies, the most ambitious of the pre–World War I experiments in scientific internationalism, to map the fractured scientific landscape that emerged from these interwar boycotts. What survived the wreckage was a strongly disciplinary model of scientific governance, eager for government funding but resistant to governmental priorities—and still more resistant to a more comprehensive vision of scientific internationalism that encompassed all countries and disciplines, despite the rhetoric of *the* scientific community.

One important exception to the reign of disciplines was the World Meteorological Organization, the most spectacular case of an independent scientific disciplinary organization trying to reconstitute itself after World War II as an intergovernmental organization and indeed as part of the newly established United Nations. The WMO is emblematic of the Faustian choice faced by many scientific organizations between official recognition and monetary support on the one hand, and autonomy and inclusive internationalism on the other.

With these two contrasting examples of mid-twentieth-century scientific internationalism in mind, I then turn to some of the earliest articulations of the novel idea of the scientific community as a kind of global village, geographically dispersed yet unified in its values, in the early years of the Cold War. The chapter concludes with some brief observations about how the scientific community in the early twenty-first century is increasingly governed neither by academies nor disciplinary organizations but rather by numbers.

3.2. The Academy of Academies: Top-Down Governance

This group photo of glum-looking gentlemen in their beards and bowlers could have been taken at almost any of the many international congresses held in the late nineteenth century: the chemists who gathered in Geneva in 1892, the astronomers who met in Paris in 1887, the mathematicians who assembled in Chicago in 1893—in short, any of the many gatherings of disciplinary elites who met in person to thrash out controversial disciplinary issues such as standardizing chemical nomenclature or cloud classification (figure 10). The worthies who posed for this photo in Wiesbaden in October 1899 certainly belonged to the most distinguished representatives of their various disciplines, but the aim of their meeting was explicitly anti-disciplinary. They were representatives of the most prestigious academies of science and letters in what they would have called "the civilized world," namely Europe and—just barely—the United States. They had been summoned to Wiesbaden in order to create their own international association of academies with the aim of combating disciplinary specialization and disciplinary autarchy—both viewed as a threat to the universalism and authority of the academies. The new organization that held its constitutive meeting in Wiesbaden in 1899 was intended to unite the sciences and the humanities in an age of splintering specialization and oversee international initiatives in all disciplines. Both the legitimacy and the efficacy of the International Association of Academies derived not only from the eminence of its member academies, representing the crème de la crème of each nation's scholars and scientists, but also from the personal reputations of its individual members—the crème de la crème de la crème.

Figure 10

From the start, the International Association of Academies spoke with a German accent. It was a "Kartel" of Germanophone academies—Vienna, Leipzig, Munich, and Göttingen—that first hatched the idea for such an association; it was the capacious German understanding of *Wissenschaft* rather than the by-then narrow English and French designation science/*science* that defined its broad ambit; and it was the old German habit of shifting the seat of its learned societies to whichever city hosted the next meeting that set the peripatetic pattern of the IAA, which wandered from Wiesbaden (1899) to London (1904) to Vienna (1907) to Rome (1910) to St. Petersburg (1913).* All

* The Versammlung Deutscher Naturforscher und Ärtzte followed this practice, as had Germany's oldest scientific academy, the Academia Naturae Curiosorum (founded 1652, later known as the Leopoldina), during its early history.

 of these Germanic influences were consequential for the IAA. Representation was not necessarily by nation but by regional academy, blurring the meaning of the word "international" in its title and weakening the IAA's diplomatic clout when it came to negotiating permission, for example, to set up observing stations in the Arctic or to survey a geodetic arc in Africa. Including the humanities as well as the sciences came naturally to the German-speaking academies and even the French, who could send emissaries from the Academy of Inscriptions and Belles Lettres in addition to those from the Academy of Sciences. But the British had no such humanistic counterpart to the Royal Society of London and hastily set in motion the creation of the British Academy, established in 1902, just in time for the IAA meeting hosted in London in 1904. Finally, although the IAA flirted with the possibility of establishing a headquarters in Brussels when the Belgian government offered international organizations highly favorable conditions, the IAA never had a permanent legal seat, greatly complicating its financial arrangements (figure 11).

But in the insouciant years between 1899 and 1914, all these seemed to be matters of secondary importance compared to the IAA's grand plans for an observer network to study atmospheric electricity or an international catalogue of scientific papers or a complete edition of the works of Leibniz. Until war was declared by the Allied and Central Powers in August 1914, no one seems to have had an inkling that the Berlin meeting planned for 1916 would never take place and that the IAA itself would be a casualty of the war. Almost all its members sided more or less wholeheartedly with their respective nations' cause, a position entirely consistent with the fin de siècle understanding

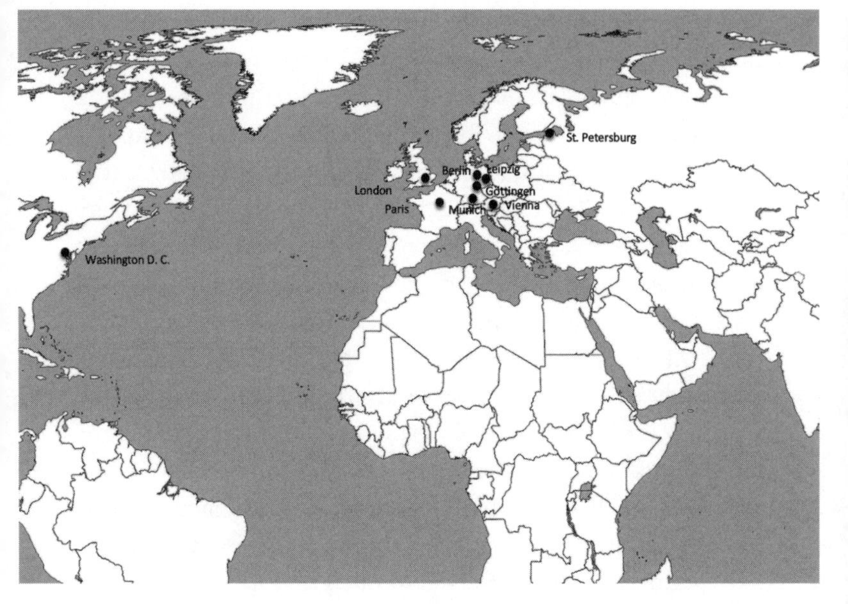

Figure 11

of scientific internationalism as a stage for national competition and mutual recognition within a shared framework of rules, much as the revived Olympic Games of the 1890s offered an arena for nationalist competition under an internationalist banner. What made continued postwar cooperation among the original members of the IAA impossible was the brutal nature of the war itself. German patriotism could be forgiven, but not German atrocities, at least not by the French and especially the Belgians, whose neutrality had been violated by the German invasion of August 1914. When in November 1918 the representatives of Allied academies met in Paris to dissolve the IAA and establish a successor organization, the International

Research Council (IRC), the French insisted that Germany and the other Central Powers be excluded. Ironically, when Brussels was chosen as the site of the IRC's first constitutive assembly in July 1919, that setting no longer symbolized inclusive internationalism but implacable hostility toward the defeated Central Powers, especially Germany. Among the Allied scientists, there was even talk about "Allied Science" being "radically different from Teutonic Science," a repudiation of the universalism of science that would have been almost unthinkable before 1914.

For their part, the defeated and excluded Germans clung even more tightly to what was left to them: their scientific and scholarly prestige, which they exploited to the hilt in the 1920s to cement bilateral relationships with both individual foreign scientists and institutions (for example, those of Japan and the Soviet Union) as a substitute for the multilateral organizations like the International Research Council that had slammed the door in their faces. In sour grapes fashion, German scientists even went so far as to derogate the importance of the international congresses and unions they had so enthusiastically participated in prior to the war, now that they were no longer invited. When in 1926 the IRC finally did relent (after much lobbying by neutral countries, such as the Dutch and the Swedes) and extend a frosty invitation to German scientists to join, the response was even frostier. Nor did the Germans and other Central Powers deign to join the successor to the IRC, the International Council of Scientific Unions, in 1931. Before the outbreak of World War II in September 1939, before even the Nuremberg Laws of 1935 made Nazi Germany a pariah, Germany had abandoned the model of nineteenth-century internationalism epitomized by the short-lived International Association of Academies.

But internationalism was not dead. Even the outcast Germans, with the strong support of the Weimar government, continued to cultivate close personal and professional relationships with individual scholars and scientists abroad. They also had at least the initial support of Albert Einstein (1879–1955), by 1919 the world's most famous scientist, who himself was not boycotted by the Allied powers because of his pacifist stance during World War I but who refused invitations to conferences that excluded his German colleagues (including the Solvay Conference), thereby boycotting the boycott. Neutral countries such as the Netherlands, Norway, Sweden, and Denmark campaigned vigorously for an end to the boycott of the scientists from the Central Powers, and offered the latter havens for collegial exchanges such as Niels Bohr's institute in Copenhagen. When Sweden joined the IRC in 1920, the Royal Swedish Academy made it clear that it would award Nobel Prizes to whomever it pleased, including Germans—a resolve they made good on multiple times in the following decade.

Moreover, the diplomatic model of internationalism conceived as gatherings of nations, each sending its delegates to congresses and collaborations, was being slowly undermined from both above and below. From above, because disciplinary unions had by 1930 become the main unit of international scientific organizations, not national academies or governments, and already in the 1920s they had rebelled against the IRC's implacable opposition to including prominent German scientists, calling the policy both anti-international and anti-scientific. From below, because scientists continued to cultivate personal ties even when official relations had been severed by war or boycotts. In 1921, for example, Dutch chemist Ernst Cohen

(1869–1944) got around the official boycott by inviting colleagues from both the Allied and Central Powers to a private conference held at his home in Utrecht. At a centenary celebration of the birth of the great French chemist Marcelin Berthelot (1827–1907) in 1927, the French chemists made a point of inviting their German colleagues not only to the official ceremony at the Collège de France but also to a private lunch at the Club de la Renaissance thereafter.

Disciplinary governance had already been the backbone of the more successful international scientific agreements and collaborations of the late nineteenth century: both the Carte du Ciel and the International Cloud Atlas were the achievements of narrowly defined disciplinary meetings, as were the standardization of nomenclature by the chemists, the botanists, and the zoologists in the same period. At least in some locales, the crystallization of disciplines—recognized identities formed by on-the-job apprenticeships (for example, as shipboard naturalists, as in the case of Charles Darwin [1809–1882], or as laboratory assistants, as in the case of Michael Faraday [1791–1867]) and later by specialized training at universities—preceded and propelled professionalization, the recognition of science as a career for which practitioners must be certified and from which they could earn a living. The reformed German universities of Göttingen, Halle, and later Berlin pioneered a model of seminar instruction that drilled small groups of advanced students in the latest findings and methods first in philology and later in fields such as physics and history. The doctoral dissertation, originally a text dictated and often written by the professor to be defended by the student, mutated into a proof of ability to conduct original, independent research as well as familiarity with the latest specialized literature and methods—whether of source criticism in history, constructing a stemma for an ancient text in philology, or making a precision measurement in physics. Seminar graduates founded, read, and contributed to a swelling tide of specialized journals—philology once again in the vanguard—that forged disciplinary reputations, standards of evidence and rigor, and accepted doctrines, often in the crucible of fiery debate.

In the course of the late-nineteenth and early-twentieth centuries, the seminar model of advanced disciplinary training set worldwide standards in science and scholarship. The students from elsewhere in Europe, the Americas, and Asia flocked to German universities to study the latest techniques in organic chemistry or in classical philology; in the United States, Johns Hopkins University (founded 1876) explicitly modeled itself on German research universities, and older universities such as Harvard soon followed suit in establishing doctoral programs. Foreign students who returned home from studies in Germany brought back more than the ability to synthesize the latest compound from coal tar or to detect a forged Latin inscription; they had been imbued with a disciplinary ethos, which lay somewhere between a religious vocation and tribal loyalty. American students in particular, many from pious backgrounds, spread the faith of *Wissenschaft* in their homeland with all the zeal of converts. The intense and demanding atmosphere of the small seminars that prepared students for a career in research in their chosen discipline forged lasting bonds of both filial piety and sibling rivalry, often at the expense of real family ties. University of Berlin historian Leopold von Ranke (1795–1886), whose seminar met daily in his home library, was incensed when some of the students asked to skip the Christmas Eve meeting in order to be with their families. Over a century later, the British biologist Peter Medawar (1915–1987) warned: "Men or women who go to the extreme length of marrying scientists should be clearly aware beforehand, instead of learning the hard way later, that their spouses are in the grip of a powerful obsession that is likely to take the first place in their lives outside the home, and probably inside too." Disciplines demanded heart-and-soul dedication, and their deeply internalized norms—first and foremost, the primacy of original, rigorous research in establishing hierarchies of recognition—continue to be transmitted in doctoral programs that directly descend from the nineteenth-century seminar-centered model.

When in 1931 the IRC was finally replaced by the International Council of Scientific Unions, it signaled the complete victory of the disciplinary unions over centralized scientific authority. From now on, internationalism would be organized at the disciplinary level, where both professional and personal ties were strongest: scientific communities, in the plural. But the Second World War and especially the Nazi murders and expulsions of Jewish scholars and scientists strained both disciplinary and personal bonds to the breaking point and beyond. The forced migration of so many intellectuals from Central Europe forged a new kind of internationalism as the émigrés and their hosts in Turkey and Palestine, in the United Kingdom and the United States were thrown together into new, long-term relationships as both university colleagues and later collaborators in military research. When the war ended in 1945, there was no more question of returning to pre-1939 arrangements in global science than there was in global politics.

3.3. Scientific Internationalism, United Nations Style: The World Meteorological Organization

The contrast between these two group photos, the one of the meteorologists who attended the Munich meeting of the International Meteorological Organization in 1891 and the other of their successors who attended the Washington, DC, meeting of the same body in 1947, goes much deeper than changes in fashion and hairstyles (figures 12 and 13). The Munich meeting represented a rejection of the organization's initial intergovernmental model of scientific internationalism, which involved nation-states sending official delegates to an international congress. Already at the very first such meeting in Vienna in 1873, participants regretted the absence of many distinguished colleagues who hadn't been chosen as their country's official delegates and listened with chagrin when other delegates who had been chosen announced that their governments considered none of the congress's resolutions—for example, the creation of a worldwide observation network—to be binding. By 1891, the International Meteorological Organization had renounced its intergovernmental status and the attendant perks of a red-carpet welcome by local dignitaries in favor of an informal gathering of the directors of various meteorological services who attended in a private capacity as scientists, not diplomats. Although the minutes of the executive council of the IMO recorded constant complaints about inadequate funds and the understaffed, overworked secretariat, the organization's technical commissions achieved considerable standardization in many aspects of meteorological observation. The *Atlas international des nuages* (1896 and subsequent editions), described in chapter two, was one product of this voluntarist model of scientific internationalism without nations.

Figure 12

Figure 13

In contrast, the 1947 Washington, DC, meeting was conducted with all the trappings of a diplomatic congress. It was sponsored by the US State Department, which supplied not only simultaneous translation (note the earphones)* and a colorful array of flags to symbolize the fifty-nine nations in attendance but also technical legal guidance for smoothing the way for the voluntarist International Meteorological Organization to become the intergovernmental World Meteorological Organization under the auspices of the newly founded United Nations. Despite the attractions of a sizable budget for pet projects like the worldwide observation network, the promised cooperation of UN member states, and the prestige of an institutional affiliation with what at the time was touted as the last, best hope for safeguarding world peace, a resolute minority of the meteorologists in attendance balked at the prospect. Would meteorologists working in countries that were not members of the United Nations be therefore excluded—for example, Switzerland, which had sheltered the IMO's secretariat during the war? What about regions that were crucial to the worldwide network of weather observations but did not belong to sovereign states—for example, Palestine (which sent two delegates to the meeting)? What was a sovereign state, anyhow (what about occupied Germany, divided into Russian, British, American, and French zones?), who decided, and what did such legal distinctions have to do with science, especially a global science like meteorology? The Irish and Norwegian delegates worried that the plenipotentiaries sent by governments might not

* The official languages of the meeting were French (the predominant language of the IMO from at least 1919 until 1945), English, Russian, and Spanish.

be meteorologists and that the smaller countries might lose their equal status with the bigger ones (no doubt a reference to the lopsided distribution of power between the UN General Assembly and the Security Council). The Rhodesian delegate proposed that any meteorological service, whether located in a territory, mandate, colony, or sovereign state, be granted equal rights of participation. Others warned that the flexibility and independence of the IMO might be encumbered by diplomatic protocols and internecine quarrels within nations—as when Egypt objected to representatives from Sudan on the grounds that Sudan was a rebel province, not an independent state. Everyone foresaw potential conflicts between political and scientific objectives: certain nations and territories were crucial to the global scope of meteorological exchanges of data, whether they chose to become members of the United Nations or not. By becoming the World Meteorological Organization, the International Meteorological Organization might ironically have to give up worldwide coverage of the weather.

Along with astronomy, meteorology had long been the global science par excellence. Since at least the late seventeenth century, observers had begun to understand the weather as a trans-local, perhaps even global phenomenon. Already in 1686 English natural philosopher Edmond Halley (who also hatched the idea for the transits of Venus observations described in chapter one) had pieced together the world system of winds from ships' logs, and throughout the eighteenth century academies exchanged weather observations in the hope of tracking how weather patterns migrated from one locale to another. In the nineteenth and twentieth centuries these global monitoring activities intensified, with the IMO leading the way in standardizing the timing, instruments, and formats of weather observations and expanding the network of regularly reporting observing stations, often in the wake of imperial expansion. After the Munich meeting of 1891 had broken with the initial intergovernmental model, the members of the IMO were directors of meteorological services, not representatives of sovereign states, and there was every motive to be as inclusive as possible: more members in more regions meant more observation stations. From this point of view, membership contingent not only on having the status of a sovereign state but a sovereign state recognized as a member of the United Nations undermined the cause of world meteorology.

Although the transformation of the IMO into the WMO was eventually approved at the Washington, DC, meeting, the predicted political problems started almost immediately thereafter. In December 1946, the UN General Assembly passed a resolution barring Spain from membership for the duration of the fascist Franco regime. In December 1947, British meteorologist Sir Nelson Johnson (1892–1954), in his capacity as president of the IMO, wrote a letter to his colleague Dr. Luis Azcárraga (1907–1988), director of the Spanish meteorological services in Madrid, to inform him that the newly created WMO was now obliged to expel all Spanish members (even those who were members of technical commissions) in compliance with the UN resolution. With eyebrow-raising chutzpah, Johnson added that he nonetheless hoped that Spanish meteorologists would conform to the new coding procedures for observations adopted at the Washington meeting. The Spanish director responded with a furious letter, insisting that the IMO was "a meeting of services and not a convention between States" and moreover that members of technical commissions were "elected individually, for their technical knowledge, without acting as representatives of their states" and that some of the Spanish members did not even reside in Spain. Other directors promptly took sides in the spat, with those from the Mediterranean countries protesting that Spain's data were essential for completing synoptic charts of the region, and the director of the Soviet Hydrometeorological Service on the contrary complaining that Johnson had gone too far in expressing his regrets in the letter to Madrid. Even before the IMO officially became the WMO in 1951, its new status as a UN agency had embroiled it in acrimonious postwar politics.

The WMO proved inventive in finding ways to soften and even evade various UN regulations and resolutions in the interests of extending its global net of observations. It played a key role in Cold War diplomacy by bringing together the United States and the USSR in a scientific cooperation during the International Geophysical Year (1957–58). During the decades of decolonization, it offered technical assistance to new nations (including Israel) on how to set up their own meteorological services as part of the UN's program of development aid—training the meteorologists from these countries in standardized WMO data formats, supplying them with standardized WMO equipment, and integrating them into the standardized global observation network protocols, most recently satellite weather and climate monitoring. All of these activities created what historian Paul Edwards has called "infrastructural globalism": a tool of governance that works by standardization and exchange of information that makes it all but impossible for nations or independent meteorological stations to secede from the WMO.

What the WMO lost in terms of inclusiveness by bowing to UN criteria for membership, it eventually regained with interest in the form of soft coercive powers to compel participation even without membership.

3.4. The Scientific Community in the Cold War: Governance against Governments

The contrast between the IMO from 1891 to 1950 and the WMO since 1951 epitomizes two contrasting models of scientific internationalism, which is all the more revealing because the subject matter, working methods, and, at least in the middle term, personnel remained largely unchanged. Both organizations were devoted to standardizing, collecting, and collating global meteorological data; both farmed out specific questions concerning, for example, cloud classification or aeronautical weather reports to technical commissions of recognized experts; both were led by directors of regional meteorological services. But while the IMO was a clubby, volunteer organization that depended on the private engagement of individuals to pay their own way to meetings, do their own translations of the proceedings, make contributions to the operating expenses of the small secretariat at their own discretion, and donate their own time and expertise to the solution of technical questions, the WMO quickly became a sprawling UN bureaucracy with a large and largely stable budget, hundreds of employees, and an imposing glass-and-steel headquarters situated among the equally imposing headquarters of other international agencies in Geneva. And while the IMO was able to subordinate nationalist politics to global disciplinary goals except during the two world wars, when almost all activity ceased, the

WMO was enmeshed in international politics from the moment of its creation as a UN agency. In the language of the sociologists, the IMO was more *Gemeinschaft* and the WMO more *Gesellschaft*. When Charles Sanders Peirce and Max Planck envisioned a trans-generational, interplanetary scientific community around 1900, it was the IMO model that they probably had in mind, so much closer to their own experience of scientific internationalism than the WMO model.

Yet when the term—and the idea—of *the* scientific community began to gain currency around 1950, it was the WMO model that was ascendant. Even if most international disciplinary unions did not become UN agencies, they also professionalized and bureaucratized. It was against this background that a new debate about the nature of the international scientific community unfolded—this time called by that name. It was a debate that involved not only scientists but also sociologists, government officials, even the CIA. In the minds of these actors, what seemed to be at stake in the decade after 1945 was not just the way in which international science should be organized but freedom itself. To do justice to these debates in their entirety would take volumes. Instead, I will focus on one episode that dramatized two contrasting and still influential views: the exchange between the Hungarian-born British chemist and philosopher Michael Polanyi (1891–1976) and the American sociologist of science Edward Shils (1910–1995) at the 1953 meeting of the Congress for Cultural Freedom in Hamburg. Was the scientific community regulated by an invisible hand, on the analogy of the free market, or by shared norms and traditions, on the analogy of a religious congregation?

The Hamburg meeting on "Science and Freedom" was a lavishly funded affair, with welcome addresses by local dignitaries, simultaneous translation in three languages, and gracious hospitality, although it is unlikely that any of the all-star cast of academic participants realized that the CIA was footing the bill.* Physicist James Franck (1882–1964), chemist Arthur Compton (1892–1962), economist Friedrich von Hayek (1899–1992), political theorist Raymond Aron (1905–1983), and many other luminaries participated, and discussions were reportedly as lively as the local hospitality was convivial. The conference had originally been convened to vaunt the freedom of Western science in contrast to repression in the Soviet bloc, and although the abuses of Lysenkoism were castigated and messages of solidarity sent to scientists behind the Iron Curtain, the delegates from the United States, Britain, France, and West Germany seemed at least as worried about the threats to scientific freedom in their own countries. The Americans wrung their hands over McCarthyism, the Britons complained about political pressure to steer research into socially useful channels, and the French warned that dependence on state funding would entangle science in bureaucratic red tape. A West German economist summarized the dilemma of postwar science: science had become more expensive and also more vital to national military and economic interests at just the moment when private funding had been decimated by the war, leaving the scientists with

* The involvement of the CIA became widely known only after 1966. On the cultural politics of the CCF, see Peter Coleman, *The Liberal Conspiracy: The Congress for Cultural Freedom and the Struggle for the Mind of Postwar Europe* (New York: Free Press, 1989).

only their governments to turn to, but at the risk of sacrificing their own research priorities and even norms like open publication and international exchange to other agendas. This became the dominant theme of the conference: how to safeguard the freedom of science from government meddling without relinquishing government generosity—basically, how to take the money and run.

Polanyi's solution was to compare the governance of the scientific community to that of the free market—in essence to assert that no governance was needed. Just as a spontaneous order emerged from the uncoordinated economic activities of individuals pursuing their own self-interest, an invisible hand coordinated the activities of individual scientists as they published their results:

> The rewards offered by the scientific world for valuable discoveries to those who first publicly claim them, and the penalties which it imposes on those who make mistaken claims, keep the work of scientists keyed up and disciplined, as the business-man's work is by the competitive pursuit of profits.

This at least was the case for pure science; Polanyi was quite willing to countenance tightly controlled hierarchical organization for applied science. Perhaps mindful of socialist critiques of individualism, Polanyi rejected all appeals to the "sanctity of the individual." Counterintuitively, he argued that public funding for science went not to individual scientists but to "scientists as members of a system of spontaneous co-ordination operating under the control of scientific opinion." The only "metaphysical" assumption he conceded was the independent existence of as-yet unknown knowledge that could only be discovered by researchers who were equally independent. The freedom of science was the freedom of the free market; the autonomy of scientists was in fact that of the science system ordered by the invisible hand of scientific consensus.

Even Friedrich von Hayek found this very Hayekian account of the scientific community too minimalist. Albeit sympathetic to Polanyi's comparison of scientific and economic activity, Hayek protested that without a collective admission of ignorance on the part of scientists—and implicitly also a collective commitment to remedying that ignorance—there would be no brake to "the totalitarian tendencies of some scientists." Others protested more vigorously. Edward Shils was particularly critical of Polanyi's free market analogy because it ignored the fact that

> all scientists together constitute a community. The collective body of scientists is more than a mere collection of separate individuals interacting with one another: it is a body of individuals bound together by a common law, such as all communities have.

If Polanyi had made academic freedom contingent upon market freedom, Shils linked it to a pluralistic society, composed of many such autonomous communities, of which science was only one.

Both Polanyi and Shils developed and modified their views in the years that followed. Polanyi never entirely gave up the free market analogy, but he placed ever more emphasis on the role of traditions of research handed down from master to apprentice and that of the "governors" of science, the elite researchers who set a discipline's standards and research agenda. For his part, Shils acknowledged the "self-maintenance and self-regulation" of science but insisted that this autonomy depended on a tradition "which lives on in memories and anecdotes as well as in scientific communications and none of which is irrelevant to keeping alive the conviction of the supreme value, within that tradition, of scientific truth." In an ironic historical reversal, scientists who had once regarded religion as the primary enemy of free inquiry, epitomized by endless retellings of Galileo's confrontation with the Catholic Church, came to regard their own community in quasi-religious terms—handed down by tradition and committed to transcendent, otherworldly values.

During the 1960s, the term "scientific community" gained a foothold first in the sociology of science, which often applied it retrospectively to any scientific collective since the seventeenth century, and gradually diffused among public policymakers, journalists, and scientists themselves.

3.5. Conclusion: The Numbers Game

Neither Polanyi nor Shils invented the term "scientific community," but their contrasting visions of what it was did chart the fault lines that divided various disciplines as to what kind of scientific community they practiced. At the Shils-ian extreme were the astronomers, whose literally cosmic subject matter and gigantic stores of data inclined them toward international cooperations remarkable for their inclusiveness and adherence to strict norms—for example, not leaking the sensational detection of gravitational waves in the constellation Virgo by the 2015 LIGO collaboration, or the first image of a black hole by the Event Horizon Telescope in the same year, until all data had been exchanged and triple-checked by scores of collaborators around the world. At the other, Polanyi-ish extreme were the buccaneering microbiologists, who had to be cajoled and threatened by both funding agencies and journal editors into archiving their genome sequencing data for the use of the discipline at large. There are still ongoing debates in the possessive biomedical disciplines about open publication of data from clinical trials.

As these examples show, the deceptive singular—*the* scientific community—hides a multitude of disciplinary diversity, with mores that run the gamut from disciplined cooperation to cutthroat competition and almost everything in between. Mutatis mutandis, the same can be said for the diversity of forms of internationalism, from reading publications of foreign colleagues in the same specialty to exchanges of postdocs between labs to full-blown, decades-long research collaborations. The scientific community materializes only in adversarial situations, conjured up by critics who take an equally monolithic view of science, whether they are from the religious right, anti-vaccination movements, or Tea Party congressmen. As one skeptic noted when the term "scientific community" was still young, it "is most often used as a strategic phrase, intended by the user to imply a large number of experts where only a few may in fact exist, or to imply unity of view where disagreement may in fact prevail."

The one common denominator amid all of this diversity is that what governance exists still takes place at the disciplinary level: it is the disciplines that set standards, both technical and ethical; it is the disciplines that select journal editors and the referees who evaluate articles and grant applications; it is the disciplines that mete out praise for achievements and blame for mistakes. The centripetal force that holds disciplines together against all of the centrifugal forces of distance, rivalry, and nationalism is ultimately expert judgment: only other specialists are qualified to award the recognition fellow specialists crave. In the language of dueling, only one's peers (and potential rivals) are capable of giving satisfaction.

However, the meaning of a "peer" in science has changed in subtle but significant ways since the late seventeenth century. The Republic of Letters defined itself as a collective of equals who coexisted in a state of sharp-elbowed freedom, the freedom to criticize without quarter or favor. But only the collective—a collective that included future as well as present members—could decide who would be admitted on terms of equality and whose judgment of merit counted; some citizens of the Republic of Letters were definitely more equal than others. Within the scientific academies of the eighteenth century, especially the more selective and prestigious ones such as the Paris Royal Academy of Sciences, ad hoc commissions evaluated manuscripts and inventions submitted for their approval, often in the capacity of official advisor to the crown. Academic publications solicited critical reviews of articles only occasionally and unsystematically; an early-nineteenth-century attempt to introduce pre-publication vetting at the Royal Society of London foundered when the two referees could not agree. Throughout the nineteenth and most of the twentieth centuries, it was journal editors, usually chosen as the doyens of their disciplines, who judged whether a submitted scientific article was fit to print or not. They resorted to external reviewers only when the burden of evaluating so many submissions became overwhelming. Granting agencies, both private and public, concentrated discretion in their directors or a small committee of hand-picked advisors and were even less inclined to seek outside assessments. Within this concertedly unsystematic system, a "peer" was not simply a fellow specialist but a recognized member of a disciplinary elite, eminent enough to be entrusted with the editorship of the field's flagship journal or decisions about which research deserved funding priority.

Peer review, in its current sense as a universal, formalized process by which all scientific articles and grant proposals are sent out to external referees for a verdict, dates only from the 1970s. It seems to have taken hold first in the United States, after Congress became increasingly nervous about large appropriations for scientific projects over which it had no direct oversight. In response to Capitol Hill grilling, the directors of the National Science Foundation and the National Institutes of Health made standardized peer review procedures the bulwark of scientific transparency and respectability. Journals and eventually private funding agencies soon followed suit, as peer review became the scientific equivalent of the Good Housekeeping Seal of Approval. A "peer" was now anyone who qualified as a referee, a foot soldier as well as a general of a discipline. The speed and uniformity with which the new standard diffused through disciplines, not only in the US but throughout the world, is a sterling example of how *the* scientific community took shape at the level of practices as a defensive response to government intrusions in scientific evaluation in the 1970s—just as the ideal of scientific community in the singular was first articulated as a defensive response to government intrusions in scientific freedom in the 1950s.

As I have argued from the outset, "community" seems a peculiarly ill-suited term to describe this fractious, competitive, dispersed, and diverse collective—if it even exists as a collective beyond the individual scientific disciplines. Yet in one crucial aspect the term was apt: as both Polanyi and Shils finally agreed, both the knowledge and the norms of scientific disciplines were transmitted by face-to-face contacts—first and foremost, between teacher and student in graduate seminars, but also within research groups, at invited colloquia and annual professional meetings. Much as in the eighteenth-century Republic of Letters and at nineteenth-century international congresses, it was these face-to-face contacts that both created and cemented shared standards, norms, and procedures, as well as the trust required to sustain international collaborations. Even two world wars could not entirely destroy these personal relationships, and after the cessation of hostilities in 1918 and 1945, it was once again these private ties that gradually mended severed lines of communication and cooperation. The scientific internationalism of the WMO contrasted sharply with that of the IMO, but without the bonds created by the former, it is very unlikely that the latter would ever have survived that contentious meeting in Washington, DC.

As in earlier centuries, these bonds are often strained by professional rivalries and national hostilities. What is new to the latter half of the twentieth century and the first decades of the twenty-first is the explosive growth of science along all dimensions: in the number of researchers (and countries where they work), in the amount of public funding, and in the number of journals and publications. Plotted over time, all these curves swoop upward, with no end in sight. It is difficult to come up with reliable statistics for how many researchers are currently at work worldwide, but a lower bound would be the 2022 OECD figures (which also include non-OECD countries China and Russia) of 3.7 million, with a steadily increasing growth rate. This means that there are more researchers currently active than at all previous epochs in the world's history put together, and the numbers keep rising.

Estimates about the number of articles in learned journals per year should also be taken with a liberal shake of salt, but the 2015 UNESCO figure of 2.5 million articles per year for peer-reviewed English-language journals only is probably an underestimate. The same source sets the number of peer-reviewed scholarly journals in English at around 28,000 in 2014. Of these, 89 percent were estimated to be in digital form in 2020 and only 10 percent in print (figure 14). The digital turn has additionally fueled growth, not least of predatory journals. Given these numbers and the inexorably upward trends, it is not surprising that older systems of evaluation, whether by editors (from the seventeenth through the early twentieth centuries) or by peer review (since the mid-twentieth century), are collapsing under the sheer weight of demand—pressure from within.

There is also pressure from without. As governments invest more and more money in research, they demand more and more evidence that the investment has paid off. However, governments lack the expertise to assess what counts as good, better, or best research by international standards, and they are wary of allowing scientists to police themselves. Here as elsewhere, this

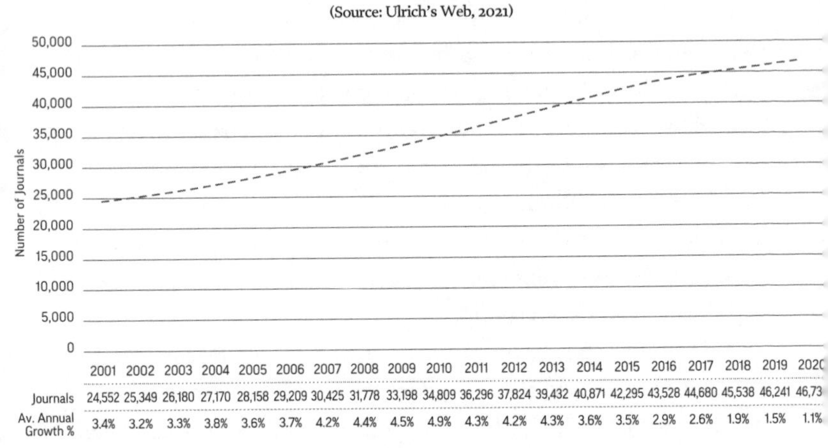

	2001	2002	2003	2004	2005	2006	2007	2008	2009	2010	2011	2012	2013	2014	2015	2016	2017	2018	2019	2020
Journals	24,552	25,349	26,180	27,170	28,158	29,209	30,425	31,778	33,198	34,809	36,296	37,824	39,432	40,871	42,295	43,528	44,680	45,538	46,241	46,73
Av. Annual Growth %	3.4%	3.2%	3.3%	3.8%	3.6%	3.7%	4.2%	4.4%	4.5%	4.9%	4.3%	4.2%	4.3%	3.6%	3.5%	2.9%	2.6%	1.9%	1.5%	1.1%

Figure 14

combination of large investment and total cluelessness makes governments turn to independent rankings, especially international rankings of universities.

They have also turned to proxy indicators, such as the Science Citation Index, introduced as a way of evaluating journal impact in the 1970s, and to New Public Management doctrines that have focused overwhelmingly on "accountability."

As the term "accountability" suggests, it originally referred to the audit practices of businesses that aimed to increase efficiency and to make the internal dealings of a firm transparent to stockholders and regulators. "Accountability" differs from responsibility in at least two respects: first, it is externally monitored by well-defined quantitative indices rather than internally by a shared ethos; and second, violations can be punished by quasi-legal sanctions. Accountability in science is increasingly taking the form of benchmarks and indicators that allegedly serve as reliable proxies for research quality and integrity and which can be mechanically calculated and thereby avoid the abuses of the old, judgment-based system, again allegedly.

From the very outset, there were warnings that the new system of proxy indicators—number of publications, journal impact factors, citation indices—might corrupt rather than promote research integrity. But the countervailing forces, particularly from government funders, as well as the internal pressures on an explosively expanding scientific system, proved irresistible. Digitalization and the rise of online journals, including many so-called predatory journals, only increased the pressure. Even if the old, judgment-based methods had worked flawlessly, how could they possibly cope with such step-changes in the sheer size of the international research system? Numerical methods seemed to be the only way to deal with big numbers.

For obvious reasons, it is difficult to get reliable information on the incidence of scientific misconduct and outright fraud. But there are some troubling indirect indicators: the number of articles retracted from journals in the natural sciences, social sciences, and humanities has increased, most for some kind of misconduct such as plagiarism. Anonymous surveys of researchers in several fields, such as biomedicine and psychology, reveal that although very few people confess to misconduct, a surprising number of respondents believe that their colleagues are guilty of it and—the key point—that honest researchers are increasingly at a competitive disadvantage.

There is even a surprising degree of diversity in definitions as to what constitutes plagiarism within the same discipline. As bad as plagiarism, cherry-picking data, or outright fabrication of results may be, there is reason to believe that other, more subtle forms of misconduct are even more damaging—and more widespread because of power imbalances. For example, under the veil of anonymity, referees regularly pressure authors to cite the referee's own publications, and journal editors anxious about their own impact factors may not only not blow the whistle but actually coerce would-be authors to increase the number of citations to articles in that journal. Perhaps most disturbing of all, surveys of researchers that ask whether such practices count as misconduct find that many respondents don't think so—and therefore presumably have no scruples about using them.

Neither the citizens of the eighteenth-century Republic of Letters nor the globe-trotting participants in nineteenth-century international congresses nor even the architects of the twentieth-century scientific community could have imagined such a situation, in which the sheer weight of numbers—of scientists, of journals, and of grants, but also of metrics to replace expert judgment with benchmarks—would dissolve the glue that held the scientific community together. The scientific community was never exactly a peaceable kingdom. But it was held together by a hunger for recognition that only other experts could confer, and, by the late nineteenth century, face-to-face relationships among its members. The former emphasized autonomy; the latter, solidarity. As we have seen, both the autonomy and solidarity of the scientific community were always fragile and yet surprisingly durable, at least at the level of disciplines. By all indicators—number of researchers, number of publications, amount of funding—science in the twenty-first century flourishes as never before in human history, and on a worldwide scale. Yet in the age of Zoom and Hirsch factors, it remains to be seen if the autonomy and solidarity of the scientific community can survive its own success.

Being in the Room Together

In Geneva, capital of internationalism since the nineteenth century, an imposing building in the sleek, shiny style favored by organizations with "world" or "international" in their names houses the International Organization for Standardization (founded 1946). Conveniently located near the airport, the ISO hosts over 800 technical commissions of emissaries sent by its 167 member nations—commissions that quietly but consequentially meet to set standards for everything from medical clinical trials to child seats to cybersecurity to carbon emissions accounting to codes for representing international time zones and currencies—almost 25,000 such standards to date. Although compliance with these standards is voluntary, governments and manufacturers ignore them at the risk of being shut out from almost any activity that reaches out across borders, from travel to trade, from communication to collaborations. The ISO has no army, no treaties, no diplomats, no United Nations representation. There is no name for the kind of power it wields, which is neither the hard power of bombs nor the soft power of prestige. It is staunchly non-governmental; indeed, it is not even intergovernmental: most of its members are the standards-setting bodies of their nations, but regions without the status of sovereign states may also participate. You have probably never heard of the ISO. Yet its decisions rule our world down to the most particular of daily particulars.

The ISO exemplifies the kind of technical and scientific internationalism so admired by champions of world projects circa 1900. Like so many of the successes of scientific internationalism since the late nineteenth century, it sets standards, thereby manufacturing universalism through uniformity. Its seat lies in a small, neutral, polyglot country. An unobtrusive permanent staff sees to the dissemination and practical realization of the new standards, as well as to preparation for the technical commissions. The real work of hammering out the detailed specification of the standards is done by the technical commissions whose members are chosen for their expertise—and who are expected to subordinate the interests of their governments, their industries, and their own political preferences to the dictates of expertise. Like the postmasters who met in Bern in 1874, the astronomers who met in Paris in 1887, the meteorologists who met in Uppsala in 1894, and the many other specialist gatherings that have met before and after to promulgate standards for their fields, the experts recruited by the ISO are engaged in an experiment in international governance for which governments may foot the bill but don't call the shots.

Against all odds, this experiment in subtracted sovereignty for shared benefits has proved remarkably durable. A ratchet effect gives standards staying power: once widely adopted, the costs of changing them or abandoning them altogether climbs steeply. Change is not impossible, as the 2018 decision by the General Conference on Weights and Measures (CGPM) to redefine the kilogram (last standardized in 1889) demonstrates. But it is glacially slow in coming (129 years in the case of the kilogram) and requires painstaking efforts to dovetail old and new standards, so as not to lose treasure troves of past data and render much material infrastructure suddenly unusable. Standards don't just flatten the world; they steady it.

Standards are both the precondition and the product of scientific collaborations. Without such standards, scientists cannot even compare their findings, much less stitch them together into a bigger picture, as astronomers did for their assigned squares of the sky in the Carte du Ciel collaboration and more recently for the image of the black hole at the center of galaxy M87 obtained by the Event Horizon Telescope collaboration—or, for that matter, as virologists all over the world did as they raced to piece together a picture of the SARS-CoV-2 virus after January 2020. But without scientific collaborations, there would be no standards. To stay with the astronomers, the expensive, perilous transit of Venus expeditions of 1761 and 1769 could not muster sufficient collaborative will to coordinate methods and instruments, with fatal consequences for the scientific value of those costly observations. Although the challenge of coordinating scientific inquiry across space in the present as well as across time began with the concerted collective empiricism of the seventeenth century, it has thus far been met only partially and sporadically.

This book has tracked the peaks and troughs of the scientific will to collaborate for the past 350-odd years. The eighteenth-century Republic of Letters, nineteenth-century internationalism, and the twentieth-century scientific community attempted, with varying degrees of success and durability, to subdue personal rivalries, overcome (or harness) national hostilities, recruit government funding but repel government meddling, balance competing research traditions, and standardize everything from names of plants to names of stars—all in the hopes of welding individual researchers into a collective that could begin to match nature's own scale. During this period, immense changes in technology, politics, society, and economics have transformed science as they have everything else. The printing press and abundant paper supplies were as essential to the operations of the Republic of Letters in the eighteenth century as steamships and telegraphs were to the workings of international science by the end of the nineteenth century. First commercial and then colonial imperialism created the nodes of globe-spanning observation networks, especially in the southern hemisphere. The Republic of Letters was as enmeshed in the court culture of its day as the Cold War scientific community was in the geopolitics of the late twentieth century. Everywhere and always, and with maddening unpredictability, governments have helped and hindered scientific collaborations that overflowed their jurisdictions: happy to bankroll them in the name of national glory, but just as happy to wreck them in the name of national security or frugality or simple indifference. The wars, revolutions, depressions, and ideologies that shook the world also jolted science. Chapter three ends in the midst of another such earthquake, the digital revolution that promises to remake science just as thoroughly as the early modern print revolution did. Science has never been an island.

Yet scientists have never ceased to imagine themselves as autonomous, if not as autarchic. The will to collaborate has always been nurtured by the vision of science and scholarship as its own polity, governed by its own rules of judging merit, awarding recognition, and ranking reputations. Despite the vast changes in every imaginable aspect of life in the past 350 years, despite the diversity of ways in which the scientific polity has been imagined and realized, recognition conferred by impartial peers—and only by them—remains at the heart of all visions of the scientific polity. Just who those peers were—Posterity? The editors of a leading journal? The members of a national academy? The Nobel Prize Committee? Anyone with a doctorate in one's field?—has evolved over time, just as the vision of the scientific polity has, from Republic of Letters to world project to scientific community. Each of these visions bears the stamp of its era; each has shown itself to be flawed: the porcupine independence of the savants of the Republic of Letters; the arrogant exclusivity of the elites, almost all European and all male, who launched scientific world projects; and the abiding contradictions of the scientific community, addicted to but wary of government funding and its attached strings. But without some such vision of a collective to give form and meaning to the will to collaborate, that will withers. This book has also been about that essential and continuing work of the imagination.

Amid all this change and variety, can any lessons be learned about when scientific collaborations succeed? Historians break out in hives when asked to draw morals from the past for the present, and for good reason: once burned, thrice shy. Nonetheless, a few patterns do stand out amid the busy background variability. First, the most effective and long-lived collaborations—though not necessarily the most spectacular—have been about establishing standards. These have also been the collaborations that proved most adroit in maneuvering around nationalist pride, restrictions, and even outright hostilities. At the euphoric meeting of the CGPM that unanimously approved the new definition of the kilogram in 2018, Iran voted with Israel, Hungary with France, China with the United States, and after the motion was carried, all the delegates rose to their feet in a thunderous ovation worthy of a full football stadium. The new kilogram lacks the glamour of the international space station, but it is far less likely to be shipwrecked by a war or diplomatic spat.

Second, governance works best at the level of disciplines. Some of the grandest scientific achievements of past decades have been interdisciplinary projects, but they require exquisite choreography among the smaller disciplinary teams. For projects that demand stamina, sacrifice, and self-control (for example, not to leak results of a Nobel-worthy discovery until all members of the collaboration have checked and certified them), the disciplinary socialization that begins in graduate school and continues in countless professional meetings and colloquia stiffens resolve.

Third, governments are fickle friends, and meddlesome ones to boot, but they have consistently been the most reliable source of sustained support for scientific projects, especially the most ambitious ones. Private patrons inconveniently die; corporations go bust; and priorities change: the donor's heirs may care more about supporting urban renewal than cutting-edge science, and the CEO may decide a tree-planting project would do more for the firm's public image than a new laboratory. Governments, especially democratic ones with short election cycles and polarized politics, also often break their promises, but their continuity of institutional commitment is still unrivaled. Their only near competitors, the world's leading research universities, may have more long-term stamina when it comes to teaching the sciences, but they, too, now depend on crucial government funding for research. As governments all over the world increase the percentage of their GDP spent on research and development, so will scientists' dependence on these sources, which almost always come with strings attached. Governments will not only decide what research will be done, whether on cures for cancer, missile defense systems, or putting humans on Mars. They will also dictate how it will be done, through an ever-denser web of regulations. He who pays the piper calls the tune.

Is there any way for science to have its cake and eat it, too, to be dependent on government funding but independent of government interference? Historically, scientists have pulled off this feat when they could leverage national rivalries to their own ends. Rivalries, not hostilities: although warring governments have been ready to shower scientists with cash to invent new weapons and perfect old ones, it has always been at an immense cost in terms of scientific autonomy, as the scientists who worked at Los Alamos to build the first atomic bombs learned the hard way. In contrast, national rivalries for international prestige harmonize beautifully with scientists' own rivalries for international recognition. Eighteenth-century scientific academies wheedled money out of monarchs eager to outdo their counterparts in Paris or Vienna, just as nineteenth-century champions of world projects piggybacked on national efforts to compensate for military defeats and economic decline by hosting the most spectacular world exposition or funding the most titanic scientific undertaking. Nowadays, nouveau riche nations such as China and the United Arab Emirates pursue international prestige through astounding architecture and ambitious space programs. And every country is keeping careful tally of its annual roster of Nobel laureates, just as it does of its Olympic gold medals, both proxies for national excellence in international competition. There is, however, a catch: prestigious science may not be useful science, much less the science needed to make discoveries and deepen understanding.

Fourth and finally, the bonds that knit the scientific collective together, however it is imagined, are sturdiest when they are reinforced by personal, face-to-face interactions. Without scientific go-betweens like Delisle and Lalande, who traveled widely and knew everybody, the academies would probably never even have answered the letters inviting them to take part in the transits of Venus expeditions, much less sent out observers to the far corners of the earth at official expense. Without the intense discussions and equally intense socializing of the first international congresses, it is most unlikely that collaborations like the Carte du Ciel or the International Cloud Atlas could ever have enlisted their disciplines' elites at all, much less over the long haul. And without the endless conversations during the coffee breaks at conferences or over dinner after invited talks, the solidarity shown by the scientific community in the face of its critics both inside and outside government would soon crumble. Behind every bland acronym for a scientific collaboration that may involve hundreds of people, there are thousands of such small-group, face-to-face encounters—smoothing disagreements, rekindling motivations, and strengthening loyalties.

The phenomena science investigates may be global, but science itself has never been. The cozy associations of the "scientific community" mask its exclusivity: someone has always been left out of those small-group parlays. Whole categories of people were and are missing, barred by reasons of class, race, gender, confession, or ethnicity. Until recently, scientific elites have resembled other elites and have been just as blind to their own blind spots. Can collaborations that rely on small-group encounters widen the circle of the group without dissolving the bonds that hold them together? Can the scientific polity be reimagined as both meritocratic and inclusive—indeed, meritocratic *because* inclusive? The jury is still out. But however composed and recomposed, scientific elites will not disappear. The symbolic economy of ranked reputations that has powered every vision of the scientific collective since the Republic of Letters is built on judgments of good, better, best.

The dinner table repartee and coffee break conversations have left only the faintest of documentary traces, mostly in private correspondence or in the occasional memoir. Official minutes and published reports hint only obliquely at the compromises reached over the fourth glass of wine, wounded egos soothed by a deft compliment, whispered alliances made against a common enemy, conciliatory toasts raised to paper over still unresolved disagreements. Yet clearly convergence was achieved in some cases, if not at the long sessions of debate recorded in the minutes, then somehow in the interstices of meetings. The historian longs for a gabby informant. How was the trick done?

Let us return to the glassy headquarters of the ISO, where technical commissions are wrangling over the standards for baby food or face masks. Members include representatives from national standards boards, relevant industries, and perhaps some consumer groups; all have been chosen for their expertise in the matter at hand. Although the interests of the represented organizations are obvious and substantial, and although the expenses of the members of technical commissions are borne by their organizations, not the ISO, the delegates are enjoined to strict neutrality. Only technical arguments for or against a proposed standard may be put forth. The situation is not unlike that of the postmasters in Bern or the astronomers in Paris or the chemists in Geneva, and neutrality seems an even more implausible imperative in the ISO technical commissions, where market share, national dominance, money, and jobs might all be at stake. What happens when a member pushes a self-interested option, as the French astronomers pushed their refracting telescopes at the 1887 Carte du Ciel conference, and art fans pushed an amateurish pastel of a stratus cloud at the 1894 meeting of the International Cloud Atlas committee? An embedded informant from the ISO offers a glimpse of what happens now and what might have happened then: "If a new delegate [to the technical commission] advances the narrow interests of [his or her employer], others will speak right up. . . . New members often need to learn about such differences. They think they are arguing a purely technical matter, and they'll put their foot in their mouths. Someone will take them aside, show them the difference." The point of this anecdote is not that the "purely technical" has fuzzy boundaries even for experts, though that is also true. It is that the possibility of consensus and collaboration even in technical matters depends on being in the room together.

ACKNOWLEDGMENTS

My first debt is to the Israeli Historical Society for the invitation to deliver the Menahem Stern Lectures, which became the core of this book. I remember the gracious hospitality of my Jerusalem hosts, especially Professor Shmuel Feiner and his wife, Rebecca, and Dr. Ts'ela Rubel, with warmest gratitude.

My home institution, the Max Planck Institute for the History of Science, Berlin, most particularly its unfailingly helpful librarians, supported the research for this book even during the worst of the pandemic. A fellowship at the National Humanities Center in North Carolina in the fall of 2021 not only advanced the project but provided me with a lively community of fellow fellows who gave me valuable comments on an early version of what became one of the book's chapters, as did the fellows of the Wissenschaftskolleg zu Berlin. The intellectual companionship kept alive by these institutions even via the pallid medium of Zoom was a boost to the spirits during the long confinement of 2020 and 2021.

Every historian relies on the institutions that preserve archives and the archivists who make them usable. In researching this book, I was aided by archivists at the Bibliothèque de l'Observatoire de Paris, Météo-France-Paris, and the Department of Manuscripts and Archives at the Uppsala University Library. Yvon Staub, archivist of the World Meteorological Organization in Geneva, welcomed me into the basement stacks, helped me track down what sources have survived, and gave me the loveliest working space I've ever enjoyed while doing archival research: a corner in the sun-flooded cafeteria of the WMO, with a glorious view of the lake below and the mountains above. Molly Ludlam-Steinke, Luise Römer, and Sina Buchholz were ingenious and indefatigable research assistants. Josephine Fenger tracked down images and permissions with perseverance and care. My hearty thanks to all.

All books build on past scholarship. But this book in particular, which aims at a synoptic overview of a vast and complex topic, owes a great deal not only to the works cited in the footnotes but also to many conversations over many years with colleagues and students in many places. They have provided me with an intellectual community worthy of the name. I am especially indebted to Peter Galison, with whom I wrote an article over a decade ago about scientific coordination that first germinated the idea for this book, and whose own subsequent work, both in print and on film, has been a never-ending source of inspiration.

Finally, my husband, Gerd Gigerenzer, has buoyed me up with his good humor and good ideas even in the darkest times. This book is dedicated to him with love and thanks.

FURTHER READING

There are many excellent studies of the development of science during the time period covered by this book, and I have drawn gratefully upon them. Readers interested in learning more about specific topics will find references in my notes to in-depth studies.

For more general background on the context in which science evolved, the following works offer helpful surveys. (Here I will only mention works available in English, although there are important literatures in other languages that this book has relied on.) Stephen Gaukroger's two monumental volumes provide a valuable overview: *The Natural and the Human: Science and the Shaping of Modernity, 1739–1841* (Oxford: Oxford University Press, 2016), and *Civilization and the Culture of Science: Science and the Shape of Modernity, 1795–1935* (Oxford: Oxford University Press, 2020). For the eighteenth-century Republic of Letters, readers are directed to Dena Goodman, *The Republic of Letters: A Cultural History of the Enlightenment* (Ithaca, NY: Cornell University Press, 1994); Anne Goldgar, *Impolite Learning: Conduct and Community in the Republic of Letters, 1680–1750* (New Haven, CT: Yale University Press, 1996); and Bruno Belhoste, *Paris Savant: Capital of Science in the Enlightenment*, trans. Susan Emanuel (New York: Oxford University Press, 2019).

The literature on the nineteenth and twentieth centuries tends toward a plethora of more specialized monographs (again, the reader is referred to the notes), but Christopher A. Bayly, *The Birth of the Modern World, 1780–1914* (Oxford: Blackwell, 2004), is an excellent jumping-off point, particularly for the impact of new technologies of transportation and communication. There has been a burst of recent studies on science and empire, of which two pioneering collections give readers an impression of the significance and richness of the topic: Simon Schaffer, Lissa Roberts, Kapil Raj, and James Delbourgo, eds., *The Brokered World: Go-Betweens and Global Intelligence, 1770–1820* (Sagamore Beach, MA: Science History Publications, 2009), and Patrick Petitjean, Catherine Jami, and Anne Marie Moulin, eds., *Science and Empires: Historical Studies about Scientific Development and European Expansion* (Dordrecht: Kluwer, 1992). To my knowledge, there is as yet no general survey on the international history of science in the twentieth century, and the tendency toward focused studies on particular disciplines (physics, biology, statistics, etc.) and periods is compounded by an emphasis on national historiographies (the United States, the Soviet Union, etc.). Although its focus is on the language of science, Michael D. Gordin, *Scientific Babel: How Science Was Done Before and After Global English* (Chicago: University of Chicago Press, 2015), is rich in insights about the international organization of science.

NOTES

INTRODUCTION

14 **usually associated with corporations, nation-states, and intergovernmental organizations:** Mark Mazower, *Governing the World* (London: Penguin, 2012).

16 **sharing data in public archives:** See, for example, The International Consortium of Investigators for Fairness in Trial Data Sharing, "Toward Fairness in Data Sharing," *New England Journal of Medicine* 375, no. 5 (2016): 405–407.

19 **before a community can come into being, it must first be imagined:** Benedict Anderson, *Imagined Communities: Reflections on the Origin and Spread of Nationalism*, rev. ed. (London: Verso, 2006); Linda Colley, *Britons: Forging the Nation 1707–1837* (London: Pimlico, 2003).

CHAPTER ONE

22 **Nor did he think he could benefit from the observations and experiments of others:** René Descartes, *Discourse on Method,* trans. Robert Stoothoff, in *The Philosophical Writings of Descartes,* trans. John Cottingham, Robert Stoothoff, and Dugald Murdoch (1637; repr., Cambridge: Cambridge University Press, 1985), 1:111–151, on 148.

25–26 **are committing the whole history of science and letters to memory:** Sigmund Jacob Apin, *Anleitung wie man Bildnüsse beruhmter und gelehrter Männer mit Nutzen sammeln und denen dagegen gemachten Einwendungen gründlich begegnen soll* (Nuremberg: Adam Jonathan Felßecker, 1728), 60. On early modern scholarly portrait collecting, see Françoise Waquet, *Respublica academia: Rituels universitaires et genres du savoir (XVIIe–XXIe siècles)* (Paris: Presses de l'Université Paris-Sorbonne, 2010), 32–37.

26 **these ghostly companions were always on their best behavior:** Thomas Babington Macaulay, "Francis Bacon," in *Literary Essays Contributed to the* Edinburgh Review (1837; repr., London: Milford, 1932), 289–410, on 291.

26 **its members aimed to produce a comparative anatomy of animals as a group:** Claude Perrault, "Projet pour les Experiences et Observations Anatomiques," in *15 janvier 1667, Procès-Verbaux*, t. 1, 22–30, Archives de l'Académie des Sciences, Paris, Registre de physique, 22 décembre 1666–avril 1668; Claude Perrault, "Projet pour la Botanique," in *15 janvier 1667, Procès-Verbaux*, t. 1, 30–38. Perrault's original manuscript, which diverges in some places from the fair hand minutes of the meeting, is preserved in the

Pochette de séance for January 15, 1667.

26 **and other such animals as came their way:** Alice Stroup, *A Company of Scientists: Botany, Patronage, and Community at the Seventeenth-Century Parisian Royal Academy of Sciences* (Berkeley: University of California Press, 1990), 39.

28 **academicians who had "eyes to see such things":** Académie Royale des Sciences, "Préface" to *Mémoires pour servir à l'histoire naturelle des animaux* (Paris: Imprimerie royale, 1671). Although even here the line between the individual and collective could be contentious, e.g., concerning the ownership of the skeletons of the dissected animals: Anita Guerrini, "Duverney's Skeletons," *Isis* 94, no. 4 (2003): 577–603.

29 **had to explicitly forbid name-calling at meetings and in print:** "L'Académie veillera exactement à ce que, dans les occasions où quelques académiciens seront d'opinions différentes, ils n'emploient aucun terme de mépris ni d'aigreur l'un contre l'autre, soit dans leurs discours, soit dans leurs écrits; et lors même qu'ils combattront les sentiments de quelques savants que ce puisse être, l'Académie les exhortera à n'en parler qu'avec ménagement." Académie Royale des Sciences, *Règlements de 1699*, p. LV, XXVI, reprinted in Léon Aucoc, *L'Institut de France. Lois, statuts et règlements concernant les anciennes Académies et l'Institut, de 1635 à 1889* (Paris: Imprimerie nationale, 1889), LXXXIV–XLX.

29 **specializing in portraits of scholars who had been the sons of prostitutes:** Sigmund Jacob Apin, *Anleitung wie man Bildnüsse beruhmter und gelehrter Männer mit Nutzen sammeln und denen dagegen gemachten Einwendungen gründlich begegnen soll*, 24.

29 **described the Republic of Letters as a bellicose state of nature:** Pierre Bayle, "Catius," in *Dictionnaire historique et critique* (Amsterdam: Reinier Leers, 1696), quoted in Françoise Waquet, *Respublica academia: Rituels universitaires et genres du savoir (XVIIe–XXIe siècles)*, 43.

30 **supplied one frequently invoked blueprint for such a collective:** Francis Bacon, *The Great Instauration and the New Atlantis*, ed., Jerry Weinberger (1626; repr., Arlington Heights, IL: Harlan Davidson, 1980), 70.

30 **can perhaps be compensated for by several people banding together:** Johann Laurentius Bausch, "Epistola invitatoria," (1652), quoted in Uwe Müller, "Die Leopoldina unter den Präsidenten Bausch, Fehr und Volckamer (1652-1693)," in *350 Jahre Leopoldina. Anspruch und*

Wirklichkeit, eds., Benno Parthier and Dietrich von Engelhardt (Halle (Saale): Deutsche Akademie der Naturforscher Leopoldina, 2002), 45–93, on 49–50.

31 **how parasitic scientific honor was upon aristocratic honor:** Michael Hunter, *The Royal Society and Its Fellows, 1660–1700: The Morphology of an Early Scientific Institution* (Chalfont St. Giles, UK: British Society for the History of Science, 1982); Michael Hunter, *Establishing the New Science: The Experience of the Early Royal Society of London* (Woodbridge, UK: Boydell Press, 1989); Roger Hahn, *The Anatomy of a Scientific Institution: The Paris Academy of Sciences, 1666–1803* (Berkeley: University of California Press, 1971).

31 **traveled in both literary and scientific circles and judged the latter to be the more vindictive:** Jean d'Alembert, "Essai sur la société des gens de lettres et des grands, sur la réputation, les mécènes et les récompenses littéraires," in *Mélanges de littérature, d'histoire et de philosophie*, nouv. ed. (Amsterdam: Z. Chatelain, 1759), 1:353.

31 **only those savants remote in both time (posterity) and space (foreigners):** On the cardinal virtue of impartiality among journal editors, see Anne Goldgar, *Impolite Learning: Conduct and Community in the Republic of Letters, 1680–1750* (New Haven, CT: Yale University Press, 1995), 99–100.

32 **ambivalence about becoming too close to foreign colleagues:** Jean d'Alembert, "Essai sur la société des gens de lettres et des grands, sur la réputation, les mécènes et les récompenses littéraires," 362. On the relationships of scientific academies to foreigners, see Lorraine Daston, "The Ideal and Reality of the Republic of Letters," *Science in Context* 4, no. 2 (1991): 367–386.

35 **dictating letter after letter containing reports of his latest observations and experiments:** Jean-Daniel Candaux, "Menu propos sur la correspondance de Charles Bonnet," in *Charles Bonnet, savant et philosophe (1720–1793)*, eds., Marino Buscaglia, René Sigrist, Jacques Trembley, and Jean Wüest (Geneva: Éditions Passé-Présent, 1994), 171–181.

35 **the closest he ever came to an international collaboration was to suggest:** Bonnet to Spallanzani, December 27, 1765, in Charles Bonnet, *Œuvres d'histoire naturelle* (Neuchâtel: Samuel Fauche, 1781), 5:10.

36 **all that would be needed would be a telescope, a reasonably good clock to time the transit:** Edmond Halley, "Methodus singularis quâ Solis Parallaxis sive

distantia à Terra, ope Veneris intra Solem conspiciendæ, tuto determinari poterit," *Philosophical Transactions* 29 (1714–1716): 454–464.

37 **for the observation of the transits of Mercury:** James E. McClellan III, *Science Reorganized: Scientific Societies in the Eighteenth Century* (New York: Columbia University Press, 1985), 203–205.

37 **even cultivated the widows of astronomers whose observations he hoped to purchase:** "Notes de Joseph-Nicolas Delisle sur sa correspondance," Bibliothèque de l'Observatoire de Paris, MS E1/3.

38 **plus the editors of various journals and monarchs such as the empress of Russia:** Delisle's list is reproduced in Harry Woolf, *The Transits of Venus* (1959; repr., New York: Arno Press, 1981), 209–211. Woolf's study remains the authoritative source for the transit expeditions of 1761 and 1769, and my account draws heavily upon it.

38 **a matter of "the honour of His Majesty and of the Nation in general":** Lord Macclesfield to the Duke of Newcastle, July 5, 1760, quoted in Harry Woolf, *The Transits of Venus*, 83.

40 **would find his observations useful "when I had quitted this life":** Quoted in Harry Woolf, *The Transits of Venus*, 122.

41 **Astronomers set to squabbling among themselves as to whose values were most trustworthy:** Harry Woolf, *The Transits of Venus*, 97–149; tables of the observers and their solar parallax values for the transits of 1761 and 1769 on 135–140 and 182–187, respectively.

43 **the weather in Zurich, Switzerland, reached Upminster, England, about five days later:** William Derham, "Tables of the Barometrical Altitudes at Zurich in Switzerland in the Year 1708. Observed by Dr. Joh. Ja. Scheuchzer, F.R.S., and at Upminster in England. Observed at the Same Time by Mr. W. Derham, F.R.S.," *Philosophical Transactions* 26 (1708-1709): 334–366.

43 **Lambert suggested that the Royal Society of London should take the lead:** Johann Heinrich Lambert, "Exposé de quelques Observations qu'on pourroit faire pour répandre du jour sur la Météorologie," *Nouveaux Mémoires de l'Académie Royale des Sciences et Belles-Lettres [de Berlin] (1771):* 60–65, quotations on 60, 61, and 62.

44 **dependent as they were on the wavering dedication of volunteer observers:** David C. Cassidy, "Meteorology in Mannheim: The Palatine Meteorological Society, 1780–1795," *Sudhoff's Archiv* 69, no. 1 (1985): 8–25, on 9–10.

44 **never traveling, and always being at home for the fixed times of observation:** Louis Cotte, *Traité de météorologie* (Paris: Imprimerie Royale, 1774), 519.

44–45 **only 14 percent were active for twenty years or more:** Gustav Hellmann, *Repertorium der deutschen Meteorologie* (Leipzig: Wilhelm Engelmann, 1883), 985–990.

45 **to volunteer the services of his court astronomer:** Harry Woolf, *The Transits of Venus*, 169–170.

47 **centralization and coordination were no substitute for proximity and personal loyalties:** Gustav Hellmann, *Repertorium der deutschen Meteorologie*, 899; David C. Cassidy, "Meteorology in Mannheim: The Palatine Meteorological Society, 1780–1795," 20, complete list of observing stations on 23–25.

47 **failed to anticipate the fragility of his own institution, the Electorate of the Palatine:** David C. Cassidy, "Meteorology in Mannheim: The Palatine Meteorological Society, 1780–1795," 19.

48 **Cotte consoled himself with thoughts of a grateful posterity:** Louis Cotte, *Mémoires sur la météorologie* (Paris: Imprimerie royale, 1788), 1:ix.

49 **funneled observations from his foreign contacts to Cotte's meteorological network:** Louis Cotte, *Traité de météorologie*, vj.

49 **was himself a member of both the Berlin and Paris academies:** Daniel Roche, *Le Siècle des lumières en province. Académies et académiciens provinciaux, 1680–1789* (Paris: Éditions de l'École des Hautes Études en Sciences Sociales, 1978), 1:321.

50 **drunken astronomers had danced the night away:** Peter Brosche, *Der Astronom der Herzogin: Leben und Werk vom Franz Xavier von Zach 1754–1832* (Frankfurt am Main: Verlag Harri Deutsch, 2001), 28, 84–96; Ken Alder, "Scientific Conventions: International Assemblies and Technical Standards from the Republic of Letters to Global Science," in *Nature Engaged: Science in Practice from the Renaissance to the Present*, eds., Mario Biagioli and Jessica Riskin (New York: Palgrave Macmillan, 2012), 19–39.

50 **a monumental project to create a new system of weights and measures based upon the meter:** On the measurement of the meter, see Ken Alder, *The Measure of All Things: The Seven-Year Odyssey and the Error That Transformed the World* (New York: Free Press, 2002).

51 **they saw absolutely no prospect of introducing it into**

public use: Johann Elert Bode, "Ueber meine Reise nach Gotha, im Jahr 1798," in *Astronomisches Jahrbuch für das Jahr 1801* (Berlin: C.F.E. Späthen, 1801), 235–239. Bode lists the astronomers in attendance on 236.

51 **chose to remember what later historians would single out as the first international scientific congress:** Peter Brosche, *Der Astronom der Herzogin: Leben und Werk vom Franz Xavier von Zach 1754–1832*, 96, 99.

52 **scientific collaborations that spanned countries and oceans seemed even more improbable:** Daniel R. Headrick, *When Information Came of Age: Technologies of Knowledge in the Age of Reason and Revolution, 1700–1850* (Oxford: Oxford University Press, 2000), 186–188.

53 **canceled by Congress after construction had already begun in Texas:** Daniel J. Kevles, "Goodbye to the SSC: On the Life and Death of the Super Superconducting Collider," *California Institute of Technology Engineering and Science* 58, no. 2 (1995): 16–25.

CHAPTER TWO

55 **world expositions that drew thousands of foreign visitors:** Claude Tapia and Jacques Taleb, "Conférences et Congrès internationaux de 1815 à 1913," *Relations internationales* no. 5 (1976): 11–35.

55 **survey of international unions in 1911 counted over 150 of them:** Paul S. Reinsch, *Public International Unions. Their Work and Organization* (Boston: Ginn and Company, 1911), 3.

56 **science in the latter half of the nineteenth century became a world project:** Markus Krajewski, *Restlosigkeit: Weltprojekte um 1900* (Frankfurt am Main: Fischer Verlag, 2006). On the number and diversity of international organizations on the eve of World War I, see John Culbert Faries, *The Rise of Internationalism* (New York: W. D. Gray, 1915).

57 **who made such maps famous in his publications:** Susan Faye Cannon, *Science in Culture: The Early Victorian Period* (New York: Science History Publications, 1978); Alexander von Humboldt, *Kosmos: Entwurf einer physischen Weltbeschreibung*, eds., Ottmar Ette and Oliver Lubrich (1845–1862; repr., Frankfurt am Main: Eichborn, 2004).

59 **Delegates from over a dozen countries met in Paris in 1872:** Peter Galison, *Einstein's Clocks, Poincaré's Maps: Empires of Time* (New York: W. W. Norton, 2003), 84–92.

59 **chemists gathered in Geneva in 1892:** Evan Hepler-Smith, "'Just

as the Structural Formula Does': Names, Diagrams, and the Structure of Organic Chemistry at the 1892 Geneva Nomenclature Conference," *Ambix* 62, no. 1 (2015): 1–28.

59 **botanists convened in Vienna in 1905:** Alphonse de Candolle, *Lois de la nomenclature botanique* (Paris: V. Masson et Fils, 1867).

59 **long-term, labor-intensive, and expensive international scientific collaborations:** Ulrich Völker, *Geschichte und Bedeutung der internationalen Erdmessung* (Munich: Bayerische Akademie der Wissenschaften, 1963).

60 **involved intricate negotiations over every detail in order to standardize the results:** Peter Galison and Lorraine Daston, "Scientific Coordination as Ethos and Epistemology," in *Instruments in Art and Science: On the Architectonics of Cultural Boundaries in the 17th Century*, eds., Helmar Schramm, Ludger Schwarte, and Jan Lazardzig (Berlin: De Gruyter, 2008), 296–333.

60 **to lower the temperature of professional quarrels by in-person sociability:** Versammlung deutscher Naturforscher und Ärtzte, *Ein Liederbuch für Naturforscher und Aerzte, als Festgabe für die Mitglieder der 41. Versammlung in Frankfurt am Main, vom 18. bis zum 24. September 1867* (Frankfurt am Main: J. D. Sauerländer's Verlag, 1867).

63 **praising the Postal Union as the harbinger of world peace:** Leonard S. Woolf, *International Government* (New York: Brentano, 1916), 197.

63 **celebrated the Postal Union as "a universal parliament":** John F. Sly, "The Genesis of the International Postal Union," in *International Conciliation. Documents for the Year 1927*, ed., Carnegie Endowment for Peace (Worcester, NY: Carnegie Endowment for International Peace, 1927), 393–446, on 396; Hugo Weithase, *Geschichte des Weltpostvereins*, 2nd rev. ed. (Strassburg: Heitz & Mündel, 1895), 52–59.

63 **letter sent from New York to Australia could travel any one of six different routes:** Joachim Helbig, *Bayerische Postgeschichte 1806–1870* (München: Helbig, 1991), 257; John F. Sly, "The Genesis of the International Postal Union," 398.

63–64 **it seemed to contemporaries tantamount to the ex nihilo creation of order out of chaos:** The treaty went into effect on July 1, 1875 and was ratified by all countries except France in October 1874; France delayed until May 3, 1875. Hubert Krains, *L'Union postale universelle: Sa foundation et son développement*

(Bern: Gustav Grunau, 1908), 48–53. It was preceded by the International Telegraph Union (established 1865), but the ITU did not enjoy the same degree of national subscription as the Postal Union did.

64 **which eventually did result in a treaty:** John F. Sly, "The Genesis of the International Postal Union," 400–401.

65 **paved the way for behind-the-scenes compromises:** Hugo Weithase, *Geschichte des Weltpostvereins*, 29–32. The countries represented at the 1863 Paris conference were Belgium, Costa Rica, Denmark, France, Great Britain, the German Hansa cities, Italy, the Netherlands, Austria, Portugal, Prussia, the Sandwich Isles, Spain, Switzerland, and the United States. Among the many exceptions eventually granted was permission to Persia to charge extra postage on the bibles sent by American missionaries that had to be transported by camel: Leonard S. Woolf, *International Government* (New York: Brentano, 1916), 204–205.

65 **moving spirits behind the Carte du Ciel and the International Cloud Atlas learned these lessons well:** Madeleine Herren, "Governmental Internationalism and the Beginning of a New World Order in the Late Nineteenth Century," in *The Mechanics of Internationalism*, eds., Martin H. Geyer and Johannes Paulmann (Oxford: Oxford University Press, 2001), 121–144.

66 **which emulsion to use on the photographic plates:** On the Carte du Ciel, see Suzanne Débarat, J. A. Eddy, Heinrich K. Eichhorn, and Arthur R. Upgren, eds., *Mapping the Sky: Past Heritage and Future Directions; Proceedings of the 133rd Symposium of the International Astronomical Union* (Dordrecht: Kluwer, 1988); Théo Weimer, *Brève histoire de la Carte du Ciel en France* (Paris: Observatoire de Paris, 1987); Jérôme Lamy, ed., *La Carte du Ciel: Histoire et actualité d'un projet scientifique international* (Paris: Observatoire de Paris, 2008).

66 **staged the Carte du Ciel with all the pomp and circumstance of a diplomatic congress:** "Soirées. Dîners à l'occasion des réunions du Comité de la Carte du Ciel," Bibliothèque de l'Observatoire de Paris, MS 1060.IV-A-2-3, Carton 25.

66 **the success of the project was "a point of honor for France":** Admiral Mouchez, Directeur de l'Observatoire de Paris, au Ministre de l'Instruction Publique, 25 April 1891, Bibliothèque de l'Observatoire de Paris, MS 1060. IV-A-2, Carton 24.

66 **"the greatest venture yet undertaken in astronomy":** Julius Scheiner, *Die Photographie der*

Gestirne (Leipzig: Wilhelm Engelmann, 1897), 311.

68 **"the authentic state of the universe visible from the earth":** Camille Flammarion, "La Photographie céleste à l'Observatoire de Paris," *Revue d'astronomie populaire* 5 (1886): 42–57, on 55.

69 **judged the costs of collaboration to be too great and declined to participate:** Pickering did, however, serve on the photometric commission of the Permanent International Committee of the Carte du Ciel: John Lankford, "The Impact of Photography on Astronomy," in *Astrophysics and Twentieth-Century Astronomy to 1950*, ed., Owen Gingerich (Cambridge: Cambridge University Press, 1984), 16–39, on 38; E. Pickering to E. Mouchez, August 14, 1889, Pickering Papers, Harvard University Archives, UAV 630.14, ser. A-9, 3-4.

69 **at the price of limiting other investigations:** Graeme L. White, "The Carte du Ciel: The Australian Connection," in *Mapping the Sky: Past Heritage and Future Directions*, eds., Suzanne Débarat, J. A. Eddy, Heinrich K. Eichhorn, and Arthur R. Upgren, 45–51, on 48; Lankford, "The Impact of Photography on Astronomy," 32, on the converse advantages to American observatories that did not participate in the Carte du Ciel.

70 **it might shame the British into paying the other half:** "Congrès 1909. Rapport au Ministre après le Congrès," Bibliothèque de l'Observatoire de Paris, MS 1060-IV-A-2, 4e Partie/ Boite 24.

70 **played a crucial diplomatic role behind the scenes:** Institut de France, Académie des sciences, *Congrès astrophotographique international* (Paris: Gauthier-Villars, 1887), pp. 74–76. The British opposed the creation of an international Bureau in Paris: A. A. Common to E. B. Mouchez, October 23, 1888, Bibliothèque de l'Observatoire de Paris, MS IV.A, B. On Mouchez's coordination with the French government, see A.M. Motais de Narbonne, "'Petite histoire' du Congrès astrophotographique de 1887," in *Mapping the Sky: Past Heritage and Future Directions*, eds., Suzanne Débarat, J. A. Eddy, Heinrich K. Eichhorn, and Arthur R. Upgren, 129–133.

71 **part of a concerted strategy after 1871 to compensate for military losses to the Germans:** Paris hosted world expositions in 1855, 1867, 1878, 1889, and 1900. On the French use of scientific and cultural eminence to boost national prestige, see Anne Rasmussen, *L'Internationale scientifique, 1890–1914* (PhD diss., École des Hautes Études en Sciences Sociales, 1995); Pascale Rabault-Feuerhahn

and Wolf Feuerhahn, eds., *La fabrique internationale de science: Les congrès scientifiques de 1865 à 1945* (Paris: CNRS Editions, 2010).

72 **disagreement as to whether, for example, the clouds of Scotland resembled those of Italy:** Robert H. Scott, "Report on the International Meteorological Conference at Munich, August 26th to September 2nd, 1891," *Quarterly Journal of the Royal Meteorological Society* 18, no. 81 (January 1892): 1–6; *Bericht über die Verhandlungen der internationalen Conferenz der Repräsentanten der Meteorologische Dienste aller Länder zu München, 26. August bis 2. September 1891* (Munich: E. Mühlthaler, 1891), 3–4.

72 **whether stable cloud forms that could be classified in genera and species existed at all:** World Meteorological Organization, *International Cloud Atlas*, vol. 1, *Manual on the Observation of Clouds and Other Meteors* (Geneva: Secretariat of the World Meteorological Organization, 1975), 11.

72 **Cloud classification had begun earlier in the century:** On the publication history of Howard's classification and slightly earlier attempts made by Jean-Baptiste Lamarck, see Gustav Hellmann, "Einleitung" to *On the Modification of Clouds*, No. 3 by Luke Howard (1803) in *Neudrucke von Schriften und Karten über Meteorologie und Erdmagnetismus*, ed., Gustav Hellmann (1894; repr., Wiesbaden: Kraus Reprint, 1969), 7–9.

73 **"publish exact representations of the form of clouds considered typical at each location":** Hugo Hildebrand Hildebrandsson, *Rapport sur les observations internationales des nuages au Comité International Météorologique* (Uppsala: Wretman, 1903), 5.

73 **they discovered that at least three designations converged:** Hugo Hildebrand Hildebrandsson, "Rapport sur la classification des nuages," in *Congrès Météorologique International, tenu à Paris du 19 au 26 septembre 1889. Procès-Verbaux Sommaires*, eds., Moureaux, Lasne and Abbé Maze (Paris: Imprimerie Nationale, 1889), 12–24, 15–16. Stratus and nimbus were used as terms in all the systems surveyed, but Hildebrandsson believed that they had different referents.

74 **designated as the International Cloud Year:** *Bericht über die Verhandlungen der internationalen Conferenz der Repräsentanten der Meteorologischen Dienste aller Länder zu München. 26 August bis 2. September 1891*, 17–18. The members of the Atlas committee were Hugo Hildebrand Hildebrandsson (Meteorological Observatory, Uppsala, Sweden),

Léon Tesseirenc de Bort (General Secretary of the Société Météorologique de France, Paris), and Abbott Lawrence Rotch (Blue Hill Observatory, Massachusetts, USA), with the right to recruit further members as needed.

74 **paintings and pastels were also included:** The nimbus (figure 13) was also represented by a painting (by Teisserenc de Bort). On the difficulties of capturing characteristic nimbus and stratus forms with photography, see Arthur W. Clayden, *Cloud Studies* (London: John Murray, 1905), 17.

75 **employed artists who worked from nature or from "good photographs":** Hugo Hildebrand Hildebrandsson, W. Köppen, and G. Neumayer, eds., *Wolken Atlas: Atlas des nuages; Cloud Atlas; Moln-Atlas* (Hamburg: Gustav Seitz, 1890).

75 **had some of the paintings redone on Weilbach's advice:** Hugo Hildebrand Hildebrandsson, *Rapport sur les observations internationales des nuages au Comité International Météorologique*, 8–9.

76 **family news sent by committee members for years afterward:** See for example A. Riggenbach to H. Hildebrandsson, Basel, October 28, 1895, Lettres à H. H. Hildebrandsson, IX. 1895–96, University of Uppsala Library, Manuscript Department, MS A281.K; also Hildebrandsson to Teisserenc de Bort, Uppsala, July 3, 1894, Bibliothèque MétéoFrance, Paris, Correspondence Hildebrandsson/Teisserenc de Bort.

76 **aim of presenting secondary as well as primary cloud forms:** *Bericht über die Verhandlungen der internationalen Conferenz der Repräsentanten der Meteorologischen Dienste aller Länder zu München. 26 August bis 2. September 1891*, 18.

76 **the Atlas committee also fixed the definitions and descriptions:** H. Hildebrandsson, A. Riggenbach, and L. Teisserenc de Bort, eds., *Atlas international des nuages. Intenationaler Wolken-Atlas. International Cloud Atlas* (Paris: Gauthier-Villars et Fils, 1896), 2.

77 **"on all sides I hear cries and demands to send the atlas!":** H. Hildebrandsson to L. Teisserenc de Bort, Uppsala, September 15, 1895, Bibliothèque MétéoFrance, Paris, Correspondence Hildebrandsson/Teisserenc de Bort.

77 **both decided that they had no authority to alter the decisions of the committee:** A. Riggenbach to H. Hildebrandsson, Basel, June 22, 1895, Lettres à H. H. Hildebrandsson, IX. 1895–96, University of Uppsala Library, Manuscript Department, MS A281.K.

77 **so as not to offend the American members attending:** H. Hildebrandsson to L. Teisserenc de Bort, Uppsala, March 5, 1893.

77 **advance copies of the chosen images were sent to tetchy German colleagues:** H. Hildebrandsson to L. Teisserenc de Bort, Uppsala, November 17, 1895.

77 **better reflected the international resolution that had brought it into existence:** H. Hildebrandsson to L. Teisserenc de Bort, Uppsala, January 4, 1896.

78 **attempted to standardize instruments, methods of measurement, and even times of observation:** On the successes and failures of the nineteenth-century international standardization movement in science and commerce, see Mark Mazower, *Governing the World* (London: Penguin, 2012), 94–115; Martin H. Geyer, "One Language for the World: The Metric System, International Coinage, Gold Standard, and the Rise of Internationalism," in *The Mechanics of Internationalism: Culture, Society, Politics*, eds., Martin H. Geyer and Johannes Paulmann, 55–92.

78 **scientific internationalism was centered on Europe and North America:** Sidney Pollard, "Free Trade, Protectionism, and the World Economy," in *The Mechanics of Internationalism: Culture, Society, Politics*, eds., Martin H. Geyer and Johannes Paulmann, 27–53, on 27–30.

79 **ultimately becoming a commission of the International Astronomical Union:** Commission 23 ("Carte du Ciel") of the International Astronomical Union existed until 1970: T. Weimer, "Naissances et developpement de la Carte du Ciel en France," in *Mapping the Sky: Past Heritage and Future Directions*, eds., Suzanne Débarat, J. A. Eddy, Heinrich K. Eichhorn, and Arthur R. Upgren, 29–32.

79 **glass photographic plates gathered dust in file cabinets:** Sean E. Urban and Thomas E. Corbin, "The Astrographic Catalogue: A Century of Work Pays Off," *Sky and Telescope* (June 1998): 40–44.

79 **it was in fact possible to calculate the proper motions of almost a million stars:** Derek Jones, "The Scientific Value of the Carte du Ciel," *Astronomy and Geophysics* 41 (2000): 5.16–5.20.

80 **most contentious of all, what would be the language of publication:** On the long history of the quest for a universal language for science, see Michael D. Gordin, *Scientific Babel: How Science Was Done Before and After Global English* (Chicago: University of Chicago Press, 2015).

81 **"astronomers of one country were not familiar with rules in use**

elsewhere": H. H. Turner, *The Great Star Map* (New York: J. Murray, 1912), 19–20.

82 **Scandalized though he was by the French practice of having the chairman move motions in meetings:** H. H. Turner, *The Great Star Map*, 18.

82 **"it is in the fostering of this feeling, much more than in the discussion of abstruse scientific questions":** Robert H. Scott, "Report on the International Meteorological Conference at Munich, August 26th to September 2nd, 1891," 1.

CHAPTER THREE

85 **an international conference in Paris on how to determine the shape of the earth:** *Verhandlungen der vom 20. Bis 29. September 1875 in Paris vereinigten Permanenten Commission der Europäische Gradmessung* (Berlin: Reimer, 1875). On the origins of the project, see J. J. Baeyer, *Das Messen auf der sphäroidischen Erdoberfläche als Erläuterung meines Entwurfes zu einer mitteleuropäischen Gradmessung* (Berlin: Reimer, 1862).

85 **he entirely subscribed to the combined ethos of precision and coordination:** Peter Galison and Lorraine Daston, "Scientific Coordination as Ethos and Epistemology," in *Instruments in Art and Science: On the Architectonics of Cultural Boundaries in the 17th Century*, eds., Helmar Schramm, Ludger Schwarte, and Jan Lazardzig (Berlin: De Gruyter, 2008), 296–333.

86 **"international solidarity" required for the scientific project to succeed:** Charles Sanders Peirce, "Measurements of Gravity at Initial Stations in America and Europe," *Coast Survey Report* (1879), reprinted in *Writings of Charles S. Peirce. A Chronological Edition*, ed., Christian J. W. Kloesel (Indianapolis: Indiana University Press, 1986), 4:79–144, on 81.

86 **"beyond this geological epoch, beyond all bounds":** Charles Sanders Peirce, "Three Logical Sentiments," in *Collected Papers of Charles Sanders Peirce*, eds., Charles Hartshorne and Paul Weiss, vol. 2, *Elements of Logic* (Cambridge, MA: Harvard University Press, 1932), 395–400, on 398.

86 **"also the inhabitants of other planets":** Max Planck, *Acht Vorlesungen über theoretische Physik: Gehalten an der Columbia University in the City of New York im Frühjahr 1909* (Leipzig: S. Hirzel, 1910), 6.

90 **The new organization that held its constitutive meeting in Wiesbaden in 1899:** The academies represented at the Wiesbaden meeting were: the Königliche Preussische Akademie der Wissenschaften zu Berlin; the Königliche Gesellschaft der

Wissenschaften zu Göttingen; the Königliche Sächsische Gesellschaft der Wissenschaften zu Leipzig; the Royal Society of London; the Académie des Sciences (Paris); the Königliche Bayerische Akademie der Wissenschaften (Munich); the Imperial Academy of St. Petersburg; the Reale Accademia die Lincei (Rome); the Kaiserliche Akademie der Wissenschaften (Vienna); and the National Academy of Sciences (Washington, DC). Invitations were issued to nine other academies. By the time of the 1913 St. Petersburg meeting, the academies of Amsterdam, Brussels, Budapest, Christiana, Copenhagen, Madrid, Stockholm, and Tokyo had also joined.

90 **from the personal reputations of its individual members:** On the origins and brief history of the International Association of Academies, see Martin Gierl, *Geschichte und Organisation: Institutionalisierung als Kommunikationsprozess am Beispiel der Wissenschaftsakademien um 1900* (Göttingen: Vandenhoek & Ruprecht, 2004); Frank Greenway, *Science International: A History of the International Council of Scientific Unions* (Cambridge: Cambridge University Press, 1996); Brigitte Schröder-Gudehus, "Les congrès scientifiques et la politique de coopération internationale des académies des sciences," *Relations internationales* no. 62 (1990): 135–148.

92 **the IAA never had a permanent legal seat, greatly complicating its financial arrangements:** On the internationalist politics of smaller countries such as Belgium and Switzerland, see Anne Rasmussen, *L'Internationale scientifique, 1890–1914* (PhD diss., École des Hautes Études en Sciences Sociales, 1995), 342–360; Madeleine Herren, "Governmental Internationalism and the Beginning of a New World Order in the Late Nineteenth Century," in *The Mechanics of Internationalism*, eds., Martin H. Geyer and Johannes Paulmann (Oxford: Oxford University Press, 2001), 121–144.

93 **offered an arena for nationalist competition under an internationalist banner:** Geert J. Somsen, "A History of Universalism: Conceptions of the Internationality of Science from the Enlightenment to the Cold War," *Minerva* 46, no. 3 (2008): 361–379, on 365–367.

94 **"Allied Science" being "radically different from Teutonic Science":** Michael Pupin to George Ellery Hale, October 20, 1917, quoted in Paul Forman, "Scientific Internationalism and Weimar Physicists: The Ideology and Its Manipulation in Germany after World War I," *Isis* 64, no. 2 (1973): 151–180, on 158, n. 18. Forman notes that there had been similar national slurs in the immediate aftermath of

the Franco-Prussian War and of course in the context of the Third Reich in Germany.

94 **went so far as to derogate the importance of the international congresses and unions:** Brigitte Schröder-Gudehus, *Deutsche Wissenschaft und internationale Zusammenarbeit, 1914–1928* (Geneva: Imprimerie Dumaret & Golay, 1966), 111–134, 213–235.

94 **Nor did the Germans and other Central Powers deign to join the successor to the IRC:** Geert J. Somsen, "Scientists of the World Unite: Socialist Internationalism and the Unity of Science," in *Pursuing the Unity of Science: Ideology and Scientific Practice from the Great War to the Cold War*, eds., Harmke Kamminga and Geert J. Somsen (London: Routledge, 2016), 82–108, on 88–92.

95 **thereby boycotting the boycott:** Michael D. Gordin, "A Century of Scientific Boycotts," *Nature* 606, no. 7912 (2022): 27–29.

95 **made it clear that it would award Nobel Prizes to whomever it pleased, including Germans:** Daniel Fauque and Robert Fox, "Binding the Wounds of War: Internationalism, National Interests, and the Order of World Science, 1919–1931," in *Blockades of the Mind: Science, Academies, and the Aftermath of the Great War*, eds., Wolfgang U. Eckart and Robert Fox, *Acta Historica Leopoldina* 78 (Halle [Saale]: Deutsche Akademie der Naturforscher Leopoldina, 2021), 41–68, on 46. Between 1920 and 1930, Nobel Prizes went to at least ten laureates from the Central Powers.

96 **but also to a private lunch at the Club de la Renaissance thereafter:** Somsen, "Scientists of the World Unite," 88; Fauque and Fox, "Binding the Wounds of War," 55–57. For the efforts of the astronomers to circumvent the boycott, see Florian Languens, "'International Science Is Bound to Win.' Eddington, Strömgren et la coopétration scientifique (1919–1922)," in *Blockades of the Mind: Science, Academies, and the Aftermath of the Great War*, eds., Wolfgang U. Eckart and Robert Fox, 27–40.

96 **science as a career for which practitioners must be certified and from which they could earn a living:** Steven Shapin, *The Scientific Life: A Moral History of a Late Modern Vocation* (Chicago: University of Chicago Press, 2008), 21–46.

96 **first in philology and later in fields such as physics and history:** On the seminar in philology, see R. Steven Turner, "Historicism, *Kritik*, and the Prussian Professoriate, 1790–1840," in *Philologie et Herméneutique au 19. siècle*, eds., M.

Bollack and H. Wiemann (Göttingen: Vandenhoek & Ruprecht, 1983), 450–489; in physics, Kathryn M. Olesko, *Physics as a Calling: Discipline and Practice in the Königsberg Seminar for Physics* (Ithaca, NY: Cornell University Press, 1991); in history, Kasper R. Eskildsen, *Modern Historiography in the Making: The German Sense of the Past, 1700–1900* (London: Bloomsbury Academic, 2022). On the origins and rise of the research university in Germany, William Clark, *Academic Charisma and the Origins of the Research University* (Chicago: University of Chicago Press, 2006); Chad Wellmon, *Organizing Enlightenment: Information Overload and the Invention of the Modern Research University* (Baltimore: Johns Hopkins University Press, 2015).

96–97 **proof of ability to conduct original, independent research as well as familiarity with the latest specialized literature and methods:** Ku-ming Kevin Chang, *The Dissertation: A Global History* (Cambridge, MA: Harvard University Press, forthcoming).

97 **older universities such as Harvard soon followed suit in establishing doctoral programs:** Emily J. Levine, *Allies and Rivals: German-American Exchange and the Rise of the Research University* (Chicago: University of Chicago Press, 2021).

97 **American students in particular, many from pious backgrounds:** Charles Rosenberg, *No Other Gods: On Science and American Social Thought*, rev. ed. (Baltimore: Johns Hopkins University Press, 1997), 135–152.

98 **was incensed when some of the students asked to skip the Christmas Eve meeting in order to be with their families:** Kasper R. Eskildsen, "Leopold Ranke, la passion de la critique et le séminaire d'histoire," in *Lieux de savoir: Espaces et communautés*, ed., Christian Jacob (Paris: Albin Michel, 2007), 462–482.

98 **"their spouses are in the grip of a powerful obsession":** Peter B. Medawar, *Advice to a Young Scientist* (New York: Harper & Row, 1979), 22. Recent focus groups conducted with scientists to investigate adherence to research norms suggest that not much has changed in expectations of devotion to discipline above all else: Melissa S. Anderson, Emily A. Ronning, Raymond DeVries, and Brian C. Martinson, "Extending the Mertonian Norms: Scientists' Subscription to Norms of Research," *Journal of Higher Education* 81, no. 3 (2010): 366–393.

99 **other delegates who had been chosen announced that their governments considered none of the congress's resolutions:** *Bericht über die Verhandlungen des*

Internationalen Meteorologen-Congresses zu Wien, 2.–16. September 1873 (Vienna: Königliche und Kaiserliche Hof- und Staatsdrückerei, 1873), 2, 6.

99 **the organization's technical commissions achieved considerable standardization in many aspects of meteorological observation:** On the history of the IMO, see World Meteorological Organization, *One Hundred Years of International Co-operation in Meteorology (1873-1973): A Historical Review* (Geneva: Secretariat of the World Meteorological Organization, 1973); Hendrik Gerrit Cannegieter, *The History of the International Meteorological Organization 1872-1951, Annalen der Meteorologie*, Neue Folge 1 (Offenbach am Main: Selbstverlag des Deutschen Wetterdiensts, 1963). Although the IMO also ceased activity during World War I, it resumed working relationships with scientists from the Central Powers in at least some of its technical commissions very soon after 1919: Giuditta Parolini, "Rebuilding International Cooperation in Meteorology after World War I: The Case of Agricultural Meteorology," in *Blockades of the Mind: Science, Academies, and the Aftermath of the Great War*, eds., Wolfgang U. Eckart and Robert Fox, 97–120. Although the minutes of the 1923 meeting of the IMO conference of directors records that "under the existing circumstances German meteorologists were prevented from attending," by the 1929 meeting in Copenhagen, twenty German meteorologists (in contrast to three from France and five from Britain) participated. Koniklijk Nederlandsch Meteorologisch Institut, *Report of the International Meteorological Conference of Directors and the Meeting of the International Meteorological Committeee at Utrecht*, September 1923 (Utrecht: Kemink & Zoon, 1924), 4, WMO Archives IMO1923. Arc; Secrétariat du Comité Météorologique International, *Procès-Verbaux des séances de la conference des directeurs du comité météorologique international, à Copenhague, septembre 1929*. Première partie (Utrecht: Kemink & Zoon, 1929), WMO Archives, IMO1929.v.1 Arc.

102 **Everyone foresaw potential conflicts between political and scientific objectives:** International Meteorological Organization, *Conférence des directeurs, Washington, 22 septembre—11 octobre 1947. Rapport final*. WMO Archives, 03-6377Arc, 28, 31–46.

102 **in the hope of tracking how weather patterns migrated from one locale to another:** Edmond Halley, "An Historical Account of the Trade Winds and Monsoons, Observable in the Seas between and near the Tropicks, with an Attempt

to Assign the Physical Cause of Said Winds," *Philosophical Transactions* 16 (1686–1692): 153–168; Theodore S. Feldman, "Late Enlightenment Meteorology," in *The Quantifying Spirit*, eds., Tore Frängsmyr, John L. Heilbron, and Robin E. Rider (Berkeley: University of California Press, 1990), 143–177.

103 **standardizing the timing, instruments, and formats of weather observations:** Cannegieter, *History of the International Meteorological Organization*, 183–190ff.; Katharine Anderson, *Predicting the Weather: Victorians and the Science of Meteorology* (Chicago: University of Chicago Press, 2005), 235–284. Colonial standardization efforts were not always successful: Philipp Lehmann, "Average Rainfall and the Play of Colors: Colonial Experience and Global Climate Data," *Studies in History and Philosophy of Science Part A* 70 (2018): 38–49.

103 **barring Spain from membership for the duration of the fascist Franco regime:** United Nations, General Assembly, December 10, 1946, Document A/241. The results of the vote by country are given at "Relations of Members of the United Nations with Spain: Resolution adopted by the General Assembly," United Nations Digital Library, accessed December 27, 2022, https://digitallibrary.un.org/record/671249?ln=en.

104 **complaining that Johnson had gone too far in expressing his regrets:** Secrétariat du Comité Météorologique International, Session of the Executive Council 1948, Oslo, August 12–17, 1948, WMO Archives, IMO.EC.1948, 144149.

104 **bringing together the United States and the USSR in a scientific cooperation:** Elena Aronova, "Geophysical Datascapes of the Cold War: Politics and Practices of the World Data Centers in the 1950s and 1960s," *Osiris* 32, no. 1 (2017): 307–327.

104 **integrating them into the standardized global observation network protocols:** World Meteorological Organization, "Space-Based Weather and Climate Extremes Monitoring (SWCEM)," Programmes, accessed December 28, 2022, https://public.wmo.int/en/programmes/wmo-space-programme/swcem.

104 **that makes it all but impossible for nations or independent meteorological stations to secede from the WMO:** Paul Edwards, "Meteorology as Infrastructural Globalism," *Osiris* 21, no. 1 (2006): 229–250.

106 **they also professionalized and bureaucratized:** The International Council of Scientific Unions (renamed the International Science Council), founded in 1931 to succeed the International

Research Council and based in Paris is a case in point: International Science Council (website), accessed December 30, 2022, https://council.science/.

108 **sacrificing their own research priorities and even norms like open publication and international exchange:** Ludwig Raiser, "State Support of Universities and Academic Freedom," in *Science and Freedom: The proceedings of a conference convened by the Congress for Cultural Freedom and held in Hamburg, on July 23rd–26th, 1953* (London: Secker &Warburg, 1955), 101–109.

109 **that could only be discovered by researchers who were equally independent:** Michael Polanyi, "Pure and Applied Science and Their Appropriate Forms of Organisation," in *Science and Freedom: The proceedings of a conference convened by the Congress for Cultural Freedom and held in Hamburg, on July 23rd–26th, 1953*, pp. 36–46, on 43–45.

109 **there would be no brake to "the totalitarian tendencies of some scientists":** Friedrich von Hayek, "Discussion," in *Science and Freedom: The proceedings of a conference convened by the Congress for Cultural Freedom and held in Hamburg, on July 23rd–26th, 1953*, p. 53.

109 **a body of individuals bound together by a common law:** Edward Shils, "Discussion," in *Science and Freedom: The proceedings of a conference convened by the Congress for Cultural Freedom and held in Hamburg, on July 23rd–26th, 1953*, p. 49.

110 **placed ever more emphasis on the role of traditions of research:** Michael Polanyi, *The Logic of Liberty: Reflections and Rejoinders* (London: Routledge, 1951), 54–56; Michael Polanyi, "The Republic of Science: Its Political and Economic Theory," *Minerva* 1, no. 1 (1962): 54–74; Michael Polanyi, *Personal Knowledge: Towards a Post-Critical Philosophy* (New York: Harper & Row, 1964), 53ff. Mary Jo Nye has argued that Polanyi's ideas of scientific community were greatly influenced by his thirteen years among the chemists in Berlin: Mary Jo Nye, *Michael Polanyi and His Generation: Origins of the Social Construction of Science* (Chicago: University of Chicago, 2011), 83.

110 **insisted that this autonomy depended on a tradition:** Edward Shils, "The Scientific Community: Thoughts After Hamburg," *Bulletin of the Atomic Scientists* 10, no. 5 (1954): 151–155, on 154.

110 **gradually diffused among public policymakers, journalists, and scientists themselves:** See for example Thomas S. Kuhn, *The Structure of Scientific Revolutions* (Chicago: University of Chicago

Press, 1962), Warren O. Hagstrom, *The Scientific Community* (New York: Basic Books, 1965), and Joseph Ben-David, *The Scientist's Role in Society: A Comparative Study* (Englewood Cliffs, NJ: Prentice-Hall, 1971). The latter attributes the term to Polanyi and Shils (3n6). The term is notably absent from key texts of the 1940s later credited with forging a postwar image of science, such as Robert K. Merton, "A Note on Science and Democracy," *Journal of Legal and Political Sociology* 1 (1942): 115–126, and Vannevar Bush, *Science, the Endless Frontier* (Washington, DC: United States Government Printing Office, 1945).

111 **until all data had been exchanged and triple-checked by scores of collaborators:** B. P. Abbott et al. (LIGO Scientific Collaboration and Virgo Collaboration), "Observation of Gravitational Waves from a Binary Black Hole Merger," *Physical Review Letters* 116, no. 6 (February 11, 2016), 061102; Shep Doeleman on behalf of the EHT Collaboration, "Focus on the First Event Horizon Telescope Results," *Astrophysical Journal Letters* (April 2019), accessed December 28, 2022, https://iopscience.iop.org/journal/2041-8205/page/Focus_on_EHT; W. Patrick McCray, "The Biggest Data of All: Making and Sharing the Digital Universe," *Osiris* 32, no. 1 (2017): 243–263; W. Patrick McCray, "Large Telescopes and the Moral Economy of Astronomy," *Social Studies of Science* 30, no. 5 (2000): 685–711. On the Event Horizon Telescope collaboration: Event Horizon Telescope (website), accessed December 28, 2022, https://eventhorizontelescope.org/; on the LIGO collaboration, Ligo Scientific Collaboration (website), accessed December 28, 2022, https://www.ligo.org/.

111 **who had to be cajoled and threatened by both funding agencies and journal editors:** Bruno J. Strasser, "The Data Deluge: Turning Private Data into Public Archives," in *Science in the Archives*, ed., Lorraine Daston (Chicago: University of Chicago Press, 2017), 185–202; see also Robert E. Kohler, "Drosophila and Evolutionary Genetics: The Moral Economy of Science," *History of Science* 29, no. 4 (1991): 335–375.

111 **ongoing debates in the possessive biomedical disciplines about open publication of data:** See for example The International Consortium of Investigators for Fairness in Trial Data Sharing, "Toward Fairness in Data Sharing," *New England Journal of Medicine* 375, no. 5 (2016): 405–407.

111 **"is most often used as a strategic phrase":** Wallace S. Sayre, "Scientists and American Science Policy," in *Sociology of Science*, eds., Bernard Barber and Walter Hirsch

(New York: Free Press of Glencoe, 1962), 596–609, on 597.

112 **often in the capacity of official advisor to the crown:** Roger Hahn, *Anatomy of an Institution: The Paris Academy of Sciences, 1666–1803* (Berkeley: University of California Press, 1971), 58–75.

112 **foundered when the two referees could not agree:** Alex Csiszar, *The Scientific Journal: Authorship and the Politics of Knowledge in the Nineteenth Century* (Chicago: University of Chicago Press, 2018), 119–158; Melinda Baldwin, "Credibility, Peer Review, and Nature, 1945–1990," *Notes and Records of the Royal Society* 69, no. 3 (2015): 337–352.

113 **the scientific equivalent of the Good Housekeeping Seal of Approval:** Melinda Baldwin, "Scientific Autonomy, Public Accountability, and the Rise of 'Peer Review' in the Cold War United States," *Isis* 109, no. 3 (2018): 538–558.

115 **3.7 million, with a steadily increasing growth rate:** OECD, "Researchers (indicator)," Research and development (R&D), accessed December 29, 2022, https://doi.org/10.1787/20ddfb0f-en.

115 **The digital turn has additionally fueled growth, not least of predatory journals:** The Association of Scientific, Technical and Medical Publishers (STM), *STM Global Brief 2021*, figures 12 and 13, accessed December 29, 2022, https://www.stm-assoc.org/2022_08_24_STM_White_Report_a4_v15.pdf; Agnes Grudniewicz et al., "Predatory Journals: No Definition, No Defence," *Nature 576, no. 7786* (2019): 210–212.

115 **are collapsing under the sheer weight of demand:** Robert E. Gropp, Scott Glisson, Stephen Gallo, and Lisa Thompson, "Peer Review: A System under Stress," *BioScience* 67, no. 5 (2017): 407–410.

116 **such as the Science Citation Index, introduced as a way of evaluating journal impact:** Eugene Garfield, "Citation Analysis as a Tool in Journal Evaluation," *Science* 178, no. 4060 (1972): 471–479.

116 **New Public Management doctrines:** Michael Power, *The Audit Society: Rituals of Verification* (Oxford: Oxford University Press, 1999).

117 **might corrupt rather than promote research integrity:** Alex Csiszar, "Gaming Metrics Before the Game: Citation and the Bureaucratic Virtuoso," in *Gaming the Metrics: Misconduct and Manipulation in Academic Research*, eds., Mario Biagioli and Alexandra Lippmann (Cambridge, MA: MIT Press, 2020), 31–42.

117 **the number of articles retracted from journals in the natural sciences, social sciences, and humanities has increased:** K. Brad Wray and Line Edslev Andersen, "Retractions in *Science*," *Scientometrics* 117, no. 3 (2018): 2009–2019; Eugenio Petrovich, "Practices and Malpractices: What the Analysis of Retractions Can Tell Us about the Research Ethos of the Humanities," *Distant Reading and Data-Driven Research in the History of Philosophy. The Blog of the DR2 Research Group of the University of Turin* (blog), uploaded on May 18, 2020, https://dr2blog.hcommons.org/2020/05/18/practices-and-malpractices-what-the-analysis-of-retractions-can-tell-us-about-the-research-ethos-of-the-humanities/.

118 **diversity in definitions as to what constitutes plagiarism within the same discipline:** Leslie K. John, George Loewenstein, and Drazen Prelec, "Measuring the Prevalence of Questionable Research Practices with Incentives for Truth-Telling," *Psychological Science* 23, no. 5 (2012): 524–532; Nannan Yi, Benoit Nemery, and Kris Dierickx, "Do Biomedical Researchers Differ in Their Perceptions of Plagiarism across Europe? Findings from an Online Survey among Leading Universities," *BMC Medical Ethics* 23, no. 1 (2022).

118 **presumably have no scruples about using them:** Mario Biagioli, Martin Kenney, Ben R. Martin, and John P. Walsh, "Academic Misconduct, Misrepresentation and Gaming: A Reassessment," *Research Policy* 48, no. 2 (2019): 401–413, offers a survey of recent literature on all of these topics for at least some of the natural and social sciences. For retractions in the humanities, see Gail Halevi, "Why Articles in Arts and Humanities Are Being Retracted?" *Publishing Research Quarterly* 36, no. 1 (2020): 55–62.

EPILOGUE

121 **its decisions rule our world down to the most particular of daily particulars:** On the history and functioning of the ISO, see Thomas A. Loya and John Boli, "Standardization in the World Polity: Technical Rationality over Power," in *Constructing World Culture: International Nongovernmental Organizations since 1875*, eds., John Boli and George M. Thomas (Stanford, CA: Stanford University Press, 1999), 169–197.

125 **maneuvering around nationalist pride, restrictions, and even outright hostilities:** Bureau International des Poids et Mesures, "Resolution 1 : Sur la révision du Système internationale d'unités," *in Conférence générale des*

poids et mesures. Comptes rendus de la 26e réunion de la CPGM (13–16 novembre 2018), 210–213. A video of the final vote at the Palais des Congrès in Versailles is available at: Bureau International des Poids et Mesures, "'On the revision of the International System of Units (SI)' and Voting. 26th CGPM," streamed live on November 17, 2018, YouTube video, 14:53, accessed December 30, 2022, https://www.youtube.com/watch?v=gimwAPQbHOw.

130 **"Someone will take them aside, show them the difference":** Quoted in Loya and Boli, "Standardization in the World Polity: Technical Rationality over Power," 186.

Columbia Global Reports is a publishing imprint from Columbia University that commissions authors to produce works of original thinking and on-site reporting from all over the world, on a wide range of topics. Our books are short—novella-length, and readable in a few hours—but ambitious. They offer new ways of looking at and understanding the major issues of our time. Most readers are curious and busy. Our books are for them.

Subscribe to our newsletter, and learn more about Columbia Global Reports at globalreports.columbia.edu.